# Basal Ganglia and Thalamus: Their Role in Cognition and Behavior

# Basal Ganglia and Thalamus: Their Role in Cognition and Behavior

Rita Moretti,
Paola Toree
and
Rodolofo M. Antonello
Editors

**Nova Biomedical Books**
*New York*

For permission to use material from this book please contact us:
Telephone 631-231-7269; Fax 631-231-8175
Web Site: http://www.novapublishers.com



**Library of Congress Cataloging-in-Publication Data**

Basal ganglia and thalamus : their role in cognition and behaviour / editors, Rita Moretti, Paola Torre, and Rodolfo M. Antonello.
p. ; cm.
Includes bibliographical references and index.
ISBN 978-1-60692-198-2 (hardcover)
1. Basal ganglia. 2. Thalamus. 3. Cognitive neuroscience. 4. Neuropsychology. I. Moretti, Rita. II. Torre, Paola, MD. III. Antonello, Rodolfo M.
[DNLM: 1. Basal Ganglia--physiology. 2. Cognition--physiology. 3. Neurobehavioral Manifestations--physiology. 4. Neuropsychology--methods. 5. Thalamus--physiology. WL 307 B29715 2009]
QP383.3.B365 2009
612.8'25--dc22

2008046715

*Published by Nova Science Publishers, Inc. ✝ New York*

# Contents

In: Basal Ganglia and Thalamus
Editors: R. Moretti, P. Torre and R. M. Antonello
ISBN 978-1-60692-198-2


*Chapter 1*

# Introduction

***Rita Moretti, Paola Torre and Rodolfo M. Antonello***

Basal ganglia are traditionally related to motor disorders; the most typical confirm of this thought has been the sentence reported in the Essay on the Shaking Palsy, by James Parkinson. His patients have been described completely from a motor perspective, but Dr. Parkinson concluded that "senses and intellect are uninjuried". More than 50 years after that book, Jean Charcot examined different patients suffering from Parkinson' Disease, and his conclusions are far from those of Parkinson; he wrote: "In general, psychic faculties are definitely impaired; at a given point, the mind becomes cloud".

Parkinson' Disease is the most popular icon of basal ganglia disruption, but not the unique or the most complete. Till Parkinson and Charcot, many Autohors have tried to describe what they find during their long-time observational periods at the bed of different patients, with acute cerebral stroke in the basal ganglia regions, or affected by cerebral tumours. The traditional view of simple motor coordinator-centre has been challenged, mainly for the anatomical *liaison* with cerebral cortex and cerebellum nuclei and cortex. These connections, organized into discrete anatomical loops, control different cerebral regions, involved in motor sequence, in behaviour or even in higher-functions, such as cognition.

The properties of neurons within the basal ganglia resemble the properties of neurons within the cortical areas subserved by the specific loop; in so far, neuronal activity within basal ganglia and cerebellar loops with areas of the motor cortex is highly correlated with motor parameters, and damage to these neurons leads to motor symptoms; neuronal activity within basal ganglia with areas of the prefrontal cortex is more related to voluntary deliberation, motivation and complex perception of the self, and the disruption of these circuits lead to higher –order deficits (Middleton and Strick, 2000).

Though these aspects, every neurologist or neuropsychologist who faces with the clinical variety of patients, admits that basal ganglia or thalamus lesions cause a multiplicity of symptom' description; and everybody knows the literature reports, where atypical aphasia has been associated with left thalamic lesions, as well as "modified anosoagnosia and hemisomatoagnosia, in addition to sensory deficits" without any mention of other cognitive alterations. It has been accepted that after thalamotomy or stimulation of these nuclei, dysphasia with dysarthria, reduced fluency, speech arrest, stuttering, hypophonia, amnesia have been reported (Bogousslavsky and Carrera, 2005).

The myriad of symptoms described in basal ganglia pathologies seem to be determined by the maintenance of many of the physiological and behavioural properties of the cortical areas that they subserve, but also by the possible influence of reward on neurons within prefrontal-basal ganglia circuits.

Moreover, it must be considered the effective role of white matter alterations, often seen in basal ganglia alteration (mainly in tumour, traumatic lesions and cerebrovascular accident), which might exert an extension or a modification of the possible confined effect due to a limited and selected lesion of one of the nuclei of the basal ganglia domain.

The uncertainty of the scenario, the fascination of the clinical observation, the real and effective implications for the patients compel us to collect some data on basal ganglia pathologies, either acute either degenerative, and their effects on cognition and behaviour. We did so, with a constant attention to Literature, to our experience, and to unusual perspectives, derived from the experience of many colleagues, who operate in different roles, tangentially bound to the clinical neurology, but strenuously bound to the patients and their life.

In: Basal Ganglia and Thalamus
Editors: R. Moretti, P. Torre and R. M. Antonello
ISBN 978-1-60692-198-2


*Chapter 2*

# Anatomy of Basal Ganglia and Thalamus

***Rita Moretti, Paola Torre, Rodolfo M. Antonello, and Antonio Bava****
* Dip. Fisiologia e Patologia, Università di Trieste, Italy

The basal ganglia consist of four main nuclei: the striatum, the globus pallidus, the subthalamic nucleus, and the substantia nigra; but other nuclei, such as the central complex of the thalamus or the pedunculopontine nucleus, also play a major role in basal ganglia functioning. The striatum, which is of telencephalic origin, consists of two macroscopic nuclei, the caudate nucleus and the putamen (Yelnik, 2002).

During ontogenesis, the caudate nucleus follows the curvature of the telencephalic vesicle, hence its curved shape. Its rostral portion, referred to as the head, is far more voluminous than the body. The tail is very small in human. The putamen and the globus pallidus arise at the junction of the diencephalon and telencephalon and do not exhibit a curved shape as does the caudate nucleus. Together they constitute the lenticular nucleus (Yelnik, 2002).

The globus pallidus is subdivided into and external (GPe) and an internal (GPi) segments.

The subthalamic nucleus (STN) arises from the lateral hypothalamic cell column of the diencephalon1 in a region located under the thalamus and above the mesencephalon. Its shape is that of a biconvex lens ($3 \times 5 \times 12$ mm in humans), which is obliquely oriented on the three anatomic planes: its longest axis is closer from the horizontal plane (20 degrees) than from the sagittal plane (35 degrees) (see data and literature in Yelnik, 2002).

The substantia nigra is a mesencephalic structure that comprises two main subdivisions, the pars compacta (SNc) and the pars reticulata (SNr). It extends between the subthalamic region, where it is located just beneath the STN, and the mesencephalon where it lies along the cerebral peduncle. It is obliquely oriented (45 degrees) from the horizontal plane (Yelnik, 2002).

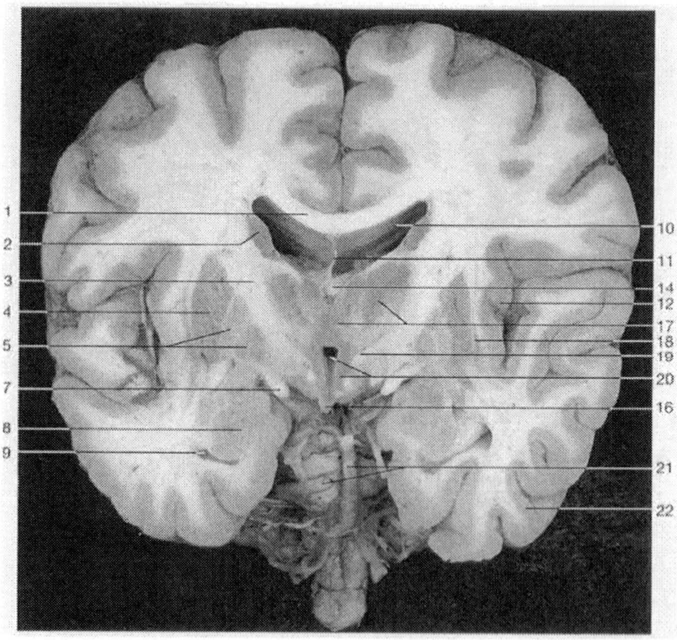


1. Corpus callosum
2. Head of the caudate nucleus
3. Internal capsule
4. Putamen
5. Globus pallidus
6. Anterior commissura
7. Optic tract
8. Amygdala
9. Temporal horn of the lateral ventricle
10. Lateral ventricle
11. Septum pellucidum
12. Insula
13. External capsule
14. Fornix
15. Optic recessum
16. Infundibulum
17. Thalamus
18. Chlaustrum
19. Lenticular ansa
20. third ventricle and hypothalmus
21. Basilar artery
22. Temporal cortex

Figure 1. A global anatomical vision of basal ganglia and thalamus.

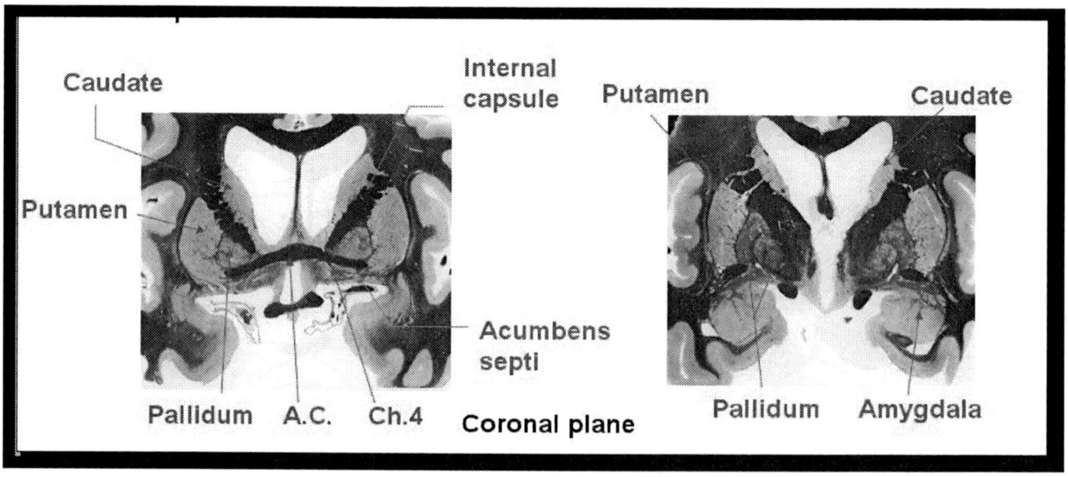


Figure 2. Reciprocal anatomical connections among basal ganglia.

The main information to the basal ganglia system arises from the cerebral cortex. The ventral part of the striatum also receives a prominent projection from the amygdale (Russchen et al., 1985; Yelnik, 2002).

Striatal information is transferred to the GPi (Yelnik et al., 1995; Yelnik, 2002), and SNr (Lynd-Balta and Haber, 1994; Yelnik, 2002) then to the thalamus (VA/VL nuclei) ( Yelnik, 2002) and from there to the frontal cortex (Barbas et al., 1991; Yelnik, 2002), supplementary motor area (SMA) for pallidal output , dorsolateral prefontal cortex for nigral output (Yelnik, 2002).

This corticocortical loop passes through the so-called direct pathway of the basal ganglia circuitry (Albin et al., 1989; Yelnik, 2002).

Another pathway, referred to as the "indirect pathway" in the model, involves the GPe, then the STN, and from there the GPi and the substantia nigra pars reticulata.

The striatum and, to a lesser degree, the globus pallidus21 and STN22–24 receive a major dopaminergic input from the substantia nigra pars compacta (Hedreen and De Long, 1991; Jan et al., 2000; Yelnik, 2002) . This control is essential to balance the activity of the direct and indirect pathways (Hirsch et al, 2000; Yelnik, 2002). The central complex of the thalamus (centre median parafascicular complex) has strong connections with the basal ganglia system in the primate (Hirsch et al, 2000; Yelnik, 2002).

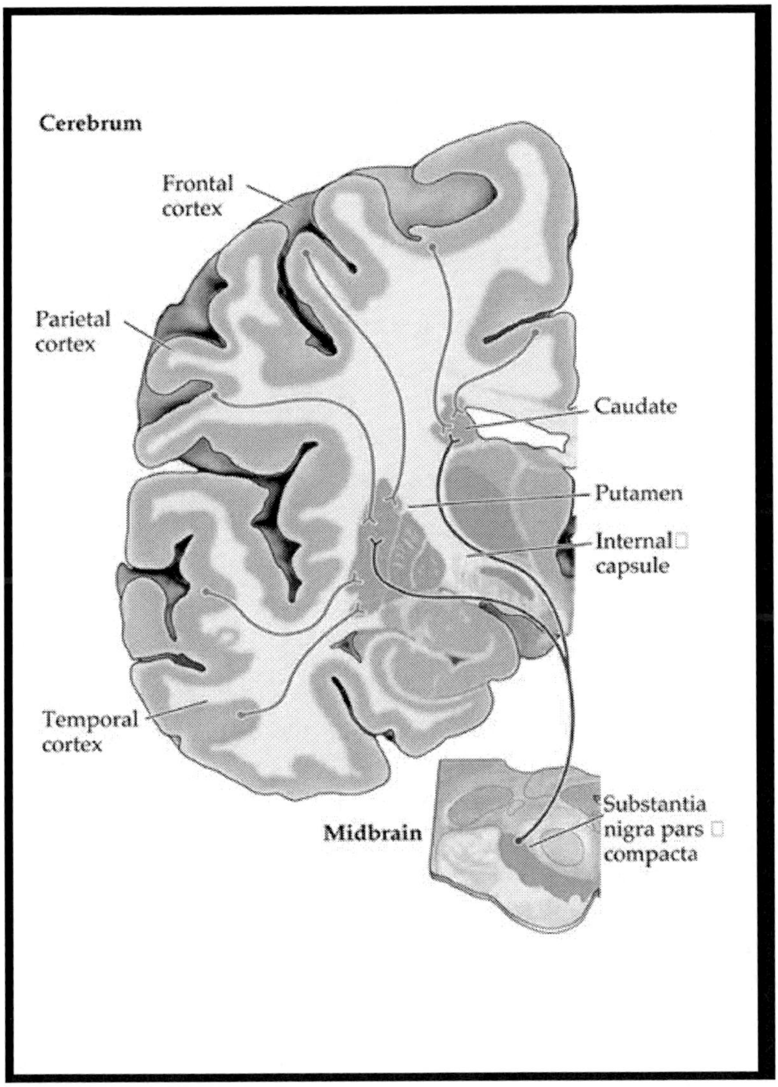


Figure 3. Inputs to the striatum.

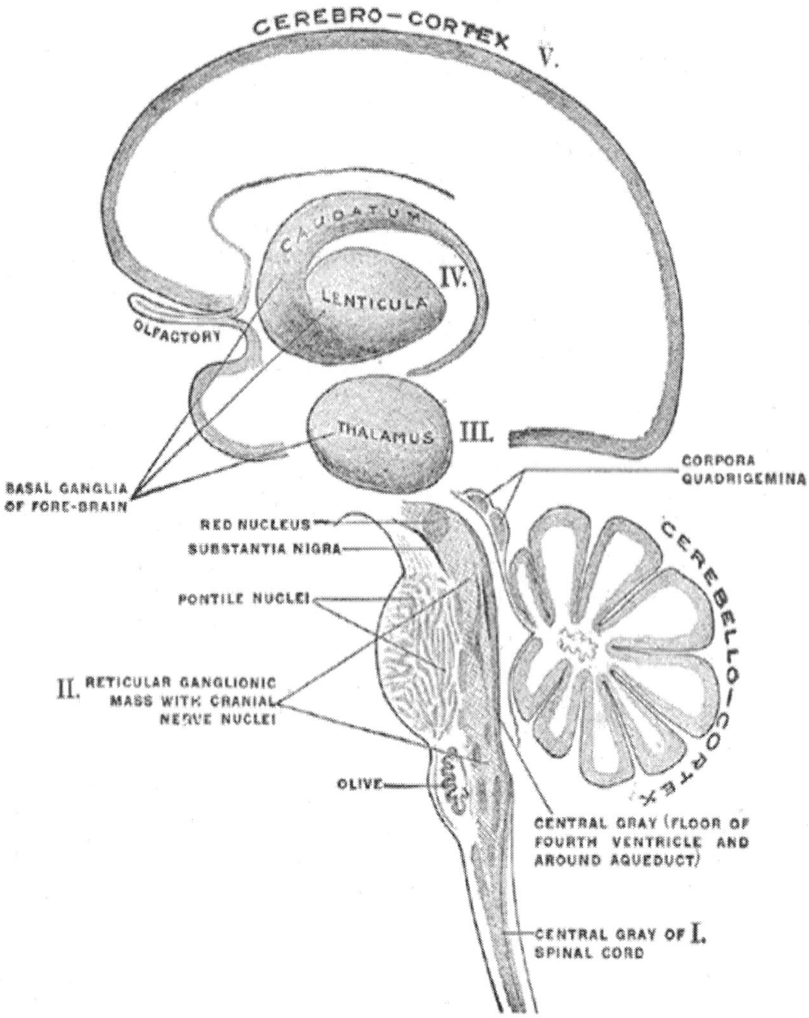


Figure 4. A functional representation of basal ganglia and thalamus connections (Mod. from Gray's Anatomy).

The GPi projects to the central part of the complex (the centre median nucleus), which projects back to the sensorimotor putamen, whereas the SNr projects to its medial part (the parafascicular nucleus), which projects back to the caudate nucleus and constitutes an important feed back on striatal activity (Nauta and Mehler, 1966; Lapper and Bolam, 1992; Yelnik, 2002).

The pedunculo-pontine nucleus has prominent projections to the basal ganglia, mainly the pars compacta of the substantia nigra and the STN. It receives its principal input from the GPi (Yelnik, 2002).

The thalamus is a nuclear complex situated in the diencephalon.

The diencephalon forms the central core of the brain and is surrounded by the hemisphere, so that only the basal surface is exposed to external view in a diamond- shaped area containing hypothalamic structures.

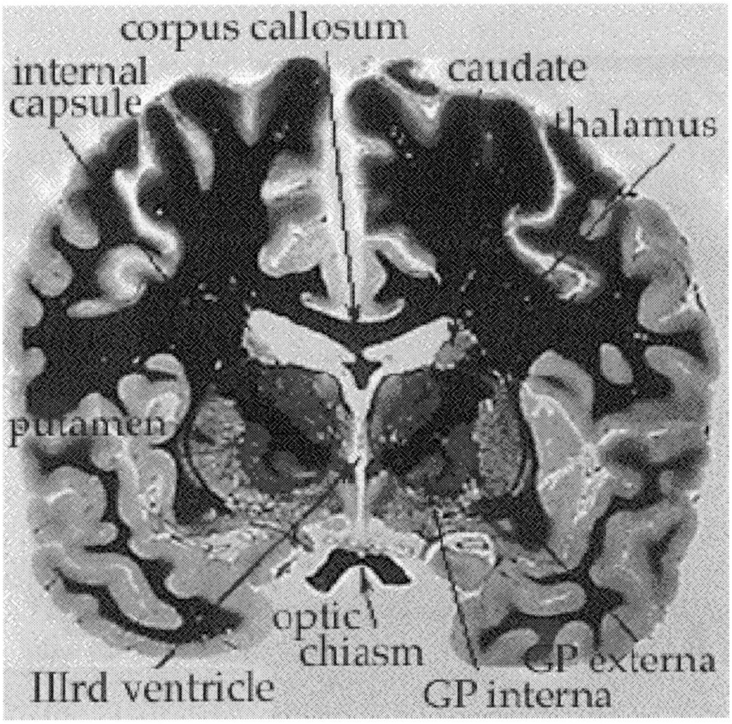


Figure 5. Reciprocal connections between thalamus and basal ganglia.

The diencephalon is located at the dorsal end of the brain stem surrounded by the internal capsule laterally and the lateral ventricles and corpus callosum superiorly. It is divided into symmetrical halves separated by the narrow third ventricle but connected by the massa intermedia. The rostrocaudal dimension of the human thalamus is about 30 mm, its height about 20 mm and its width about 20 mm. It has been estimated that there are about 10 million thalamic neurons in each hemisphere (Herrero et al., 2002).

The human thalamus is divided into 50–60 nuclei. The names of most of these nuclei are derived from their geographical locations within the thalamus.

The thalamus is made up of four parts:

1) the hypothalamus
2) the epithalamus
3) the ventral thalamus
4) the dorsal thalamus

The epithalamus belongs to the same prosomere as the dorsal thalamus; it comprises the paraventricular complex, the habenular complex and the pretectal group.

The dorsal thalamus has two regions: the allothalamic region and the isothalamic region.

The allothalamic region can be divided into:

1. The paraventricular region or midline nuclei (including the massa intermedia) (receiving afferents from the amygdala)

2. The centre-median-parafascicular complex (isolated by a continuous capsule), which appears to be a major element in the basal ganglia system (Herrero et al., 2002)
3. The intralaminar region, which includes the nuclei within the lamina medialis

The isothalamic region constitutes the bulk of the thalamus. It has "bushy" neurons and "microneurons", and most of its connections are directed to the cortical areas receiving an important mass of corticothalamic axons (in primates, much more important than the thalamocortical projection) (Percheron et al., 1996; Herrero et al., 2002). The internal and superior laminae divide the isothalamus into several subregions. The internal medullary lamina divides the principal thalamic nuclei into two major parts: a medial group on one side and a ventrolateral group on the other side of the internal medullary lamina. The anterior part of the internal medullary lamina splits into two lamellae, which surround the anterior nuclei. The cell groups within the internal medullary lamina in the caudal part are the named intralaminar nuclei (allothalamic region) (Jones, 1997; Herrero et al., 2002).

Thalamic relay neurons receive excitatory glutamatergic input from peripheral sensory pathways (Blomqvist et al., 1996; Herrero et al, 2002). Most of the thalamic nuclei project to one or few well-defined cortical areas, with the exceptions of the reticular nucleus and the intralaminar nuclei, which project diffusely to many areas of the cortex and have been defined as "nonspecific" thalamic nuclei (Jones, 1985; Herrero et al., 2002).

The centromedian and parafascicular nuclei (nucleus centralis medius and nucleus centralis parafascicularis, respectively) also project back to the striatum (Percheron et al., 1994; Herrero et al., 2002).

Moreover, thalamocortical interconnections may not be simple one-to-one connections between single thalamic nuclei and single cortical areas (Percheron et al., 1996; Herrero et al., 2002): multiple cortical areas receive afferents from a single thalamic nuclei and send information back to different thalamic nuclei (Guillery, 1995; Herrero et al., 2002).

In addition to this, the corticofugal projection provides positive feedback to the "correct" input, while at the same time suppressing irrelevant information: a given thalamic neuron receives excitation from a small cortical area that shares the same stimulus selectivity ("egocentric selection") (Herrero et al., 2002).

Each anatomically defined thalamic nucleus has its own characteristic afferent and efferent connections (Percheron et al., 1996; Herrero et al., 2002): this defines its function.

Two distinct types of thalamic nucleus are proposed on the basis of their afferent fibres from ascending pathways and/or from the cerebral cortex:

1. First-order nuclei, with a modulatory function, receive primary afferent fibres from ascending pathways and receive corticothalamic afferents from cortical layer 6, which sends branches to the reticular nucleus.
2. Higher order nuclei receive primary afferents from pyramidal neurons of the cortical layer 5 lacking a branch to the reticular nucleus; its function is concerned with transmitting information about the output of one cortical area to another cortical area (playing a part in cortico-cortical connections on higher cortical functions) (Percheron et al., 1996; Herrero et al., 2002).

The nucleus perithalamicus, also called reticular nucleus, which surrounds much of the thalamus like an eggshell, contains inhibitory GABAergic neurons, which exert weak or strong inhibitory connections with significantly different responses on thalamic neurons (Cox et al., 1997; Herrero et al., 2002).

This thalamic nucleus does not project to the cortex, however, receives collaterals from thalamocortical and corticothalamic axons. It is in an excellent position to monitor the activity in the corticothalamic and thalamocortical channels and to transmit information back to the thalamus and the midbrain.

Then the reticular nucleus controls the activity of thalamocortical channels, which is of fundamental importance for the two-way thalamocortical and corticothalamic connections (Adams et al., 1997; Herrero et al., 2002).

The intralaminar nuclei have been included in the "non-specific ascending reticular activating system" projecting to extensive areas of the cerebral cortex the inputs received from the brain stem reticular formation (Percheron et al., 1991; Herrero et al., 2002).

These nuclei have been functionally associated with attention, arousal and consciousness. The nuclei are divided into a rostral group and a central-caudal group. By way of nonselective inputs, mainly from the cholinergic brain stem nuclei as pedunculopontine nucleus and laterodorsal nucleus, the central-caudal group could modify the activity of the basal ganglia-thalamocortical loops (Percheron et al., 1991; Herrero et al., 2002).

The isothalamus is divided in different main regions.

## Regio Superior

The nucleus anterior receives projections from the mammillary body (mammillothalamic tract) and projects to the cingulate cortex belonging to the Papez circuit, the neural circuit for emotion (Papez, 1937; Herrero et al., 2002).

## Super-Regio Medioposterior

### Regio Medialis

The nucleus medialis receives a large projection from the rhinal and dorsolateral prefrontal cortices, amygdala, ventral globus pallidus and nigral projections. It projects to the orbital, medial and dorsal prefrontal cortex, and shows strong degeneration after frontal leucotomy. The nucleus shows an intensive pattern of cannabinoid receptors, as many of the regions are associated with higher functions. It has been implicated in cognition, recognition memory and habituation, in olfaction, in the wake–sleep cycle, in respiration [3] and in other vegetative and endocrine circadian activities (Pakkenberg, 1993; Pepin et al., 1993; Churchill and Kalivas, 1999; Glass et al., 1997; Ogawa et al., 1999; Akopyan et al., 2000; Barbas, 2000; Ongur and Price, 2000; Herrero et al., 2002).

### Regio Posterior

The pulvinar-lateral posterior complex, which has reciprocal connections with the associated parieto- occipito-temporal cortical areas, is involved in visual and language functions. The pulvinar has a role in selective visual attention. It has four areas (medial, central, lateral and lateral shell) demarcated by acetylcholinesterase and cytochrome oxidase (Herrero et al., 2002).

## Super- Regio Basalis

### Regio Basalis

The *nucleus suprageniculatus* is located posteromedial to the nucleus ventralis caudalis lateralis (lemniscal thalamus), ventral to the regio posterior (pulvinar) and the regio ventralis (allothalamus), lateral to the nucleus centralis parafascicularis and dorsal to the nucleus geniculatus medialis. The nucleus is the relay for pain and temperature sensation, including visceral information. The spinothalamic tract arises from cells in lamina I and lamina V of the dorsal horn and projects to the nucleus suprageniculatus of the thalamus [161] establishing simple axodendritic synapses (Craig et al., 1994; Foreman, 1999; Herrero et al., 2002).

## Super-Regio Inferolateralis

### Regio Geniculata

The *GM (nucleus geniculatus medialis)* is situated lateral to the midbrain tegmentum in the caudal region of the diencephalon rostral to the pulvinar (Herrero et al., 2002).

It has three parts:

1. A medial part with magnocellular neurons which receive somatosensory, vestibular and auditory afferents.
2. A dorsal part, a small nucleus, receiving visual and auditory inputs.
3. A ventral part with parvocellular neurons which receive only auditory signals.

The brachium from the inferior colliculus enters the nucleus in its posteromedial aspect. The acoustic radiation begins in the anterolateral border of the nucleus and curve in an anterior direction through the white matter within the transverse temporal gyrus to reach the primary auditory cortex (Herrero et al., 2002).

The *GL (nucleus geniculatus lateralis)* is positioned inferior to the pulvinar and lateral to the nucleus geniculatus medialis. In sagittal MR images the nucleus lies anterior and superior to the hippocampal formation and in coronal MR images the nucleus is situated between the pulvinar and the cerebral crus (the corticospinal fibres in the anterior part of the midbrain). The retinotopy is easily studied by functional magnetic resonance imaging.

The optic radiation begins in the posterior half of the lateral side of the nucleus and curves over the temporal horn of the lateral ventricle to reach the primary visual cortex. Both the nucleus and the optic radiation can be analysed by MRI for the interpretation of the brain lesions (Craig et al., 1994; Foreman, 1999; Chen et al., 1999; Chen and Wu, 2001; Herrero et al., 2002).

# Regio Lateralis

## Subregio Arcuata

The *gustatory thalamus*, just rostral to the lemniscal thalamus, is the relay for gustatory sensation receiving the inputs from the rostral part of the lemniscus medialis (trigeminal pathway).

The posterior border of the ventral posterolateral nucleus (nucleus ventralis caudalis lateralis) leading to the pulvinar is indicated by a triangular area (Wernicke's triangular area).

The nucleus of the subregio caudalis lemniscal thalamus is found lateral to the internal medullary lamina, caudal to the ventral intermediate nucleus (subregio intermedia) and rostral to the pulvinar. Laterally it is surrounded by the nucleus reticularis (nucleus perithalamicus) and the external medullary lamina, and dorsally by the nucleus posterior. The lemniscus medialis enters this thalamic nucleus on its inferomedial aspect at its border with the pulvinar. The nucleus has a role as a somatosensory information modulator (Herrero et al., 2002). A three-link neuronal chain forms the lemniscus medialis system, connecting the mechanosensory and propioceptive receptors neurons of the extremities, trunk, neck and back of the head with the primary somatosensory cortex throughout the thalamus using glutamate as a transmitter (Morrow and Casey, 1992; Herrero et al., 2002).

The *motor thalamus* can be identified as the nuclei that transfer information from the substantia nigra (sub-regio dorsalis), the globus pallidus (subregio oralis) and the cerebellum (subregio intermedia) to the prefrontal, supplementary, premotor, motor and somatosensory areas of the cerebral cortex (Herrero et al., 2002). Notwithstanding these segregated pathways, there are overlapping projections to the pre-supplementary motor area, where there are interdigitating foci of pallidal and cerebellar territories (Herrero et al., 2002). All the thalamic motor neurons become active before the onset of movement. In the nucleus a "voluntary unit" has been described, a cluster of neurons that are linked to rhythmic activity. These neurons are activated only by voluntary movements of the contralateral limb and not by passive movements. The motor thalamus receives projections from the nucleus perithalamicus in a specific diffuse manner, which can also contribute to the transfer of motor information ( Raeva, 1986; Percheron et al., 1996; Sakai et al., 2000; Herrero et al., 2002).

In: Basal Ganglia and Thalamus
Editors: R. Moretti, P. Torre and R. M. Antonello
ISBN 978-1-60692-198-2


*Chapter 3*

# Anatomy of White Matter

***Rita Moretti, Paola Torre and Rodolfo M. Antonello***

It has become apparent that all behaviors are subserved by distributed neural systems that comprise anatomic regions, or nodes, each displaying unique architectural properties, distributed geographically throughout cortical and subcortical areas of the nervous system, and linked anatomically and functionally in a precise and unique manner (Nauta, 1964; Luria 1966; Schmahmann and Pandya, 2005).

From an anatomical perspective, the neurons within a cortical area give rise to three distinct categories of fiber system that can be distinguished within the white matter immediately beneath the gyrus. These are association fibers, striatal fibers, and sub-cortically directed fibers (Schmahmann and Pandya, 2005).

The most fundamental association fibers of the frontal areas are the followings:

the *superior longitudinal fasciculus (SLF),* which is composed by:

1) SLF I lies medially situated in the white matter of the superior parietal lobule and the superior frontal gyrus. It is thought to play a role in the regulation of higher aspects of motor behavior that require information about body part location, and may contribute to the initiation of motor activity.
2) SLF II is more laterally situated, and occupies a position in the central core of the hemisphere white matter, lateral to the corona radiata and above the Sylvian fissure. It is thought to serve as the conduit for the neural system subserving visual awareness, the maintenance of attention, and engagement in the environment, and provides a means whereby the prefrontal cortex can regulate the focusing of attention within different parts of space.
3) SLF III is further lateral and ventral, and is located in the white matter of the parietal and frontal operculum. It provides the ventral premotor region and pars opercularis with higher order somatosensory input, may be crucial for monitoring oro-facial and hand actions (Schmahmann and Pandya, 2005).

*The arcuate fasciculus* runs first in the white matter of the superior temporal gyrus, and then in the white matter deep to the upper shoulder of the Sylvian fissure as it courses towards the frontal lobe. By linking the caudal temporal lobe with the frontal cortex dorsolateral prefrontal cortex it may be viewed as an auditory spatial bundle, important for the spatial attributes of acoustic stimuli and auditory related processing (Schmahmann and Pandya, 2005).

*The extreme capsule* is situated between the claustrum and the insular cortex caudally, and between the claustrum and the orbital frontal cortex rostrally. The extreme capsule is the principal association pathway linking the middle superior temporal region with the caudal parts of the orbital cortex and the ventral-lateral prefrontal cortex, including area 45 (Schmahmann and Pandya, 2005).

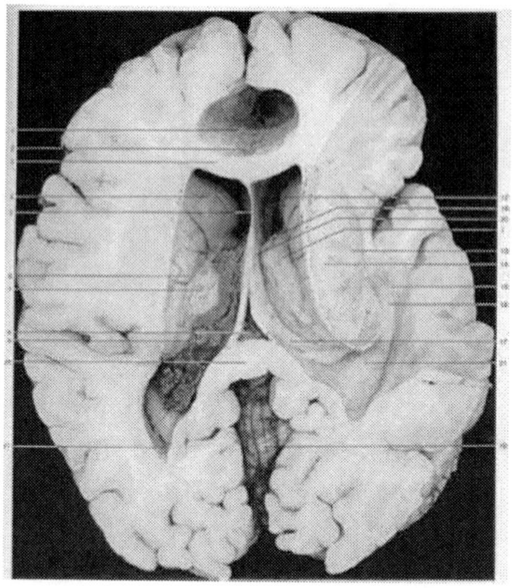

1 Stria longitudinalis lateralis;
2 Stria longitudinalis medialis;
3 Corpus callosum genu
4 Head of caudate nucleus; 5 Pellucidum septum; 6 Stria terminalis
7 Thalamus (lamina affixa); 8 Third ventricle-choroideum plexus;
9 Lateral ventricle-choroideum plexus
10 Corpus callosum splenium; 11 Hoccipital horn of lateral ventricle;
12 Frontal horn of lateral ventricle;
13 Putamen ; 14 Interal capsule; 15 Temporal horn of lateral ventricle; 16 Hippocampus foot
17 Fornix;
18 Vermis;
19 Monro forum;
20 Right fornix;
21 Collateral Prominence

Figure 6. An anatomical overview of subcortical structures.

*The middle longitudinal fasciculus* is situated within the white matter of the caudal inferior parietal lobule and extends into the white matter of the superior temporal gyrus. It links a number of high level associative and paralimbic cortical areas including the inferior parietal lobule, caudal cingulate gyrus, parahippocampal gyrus, and prefrontal cortex (Schmahmann and Pandya, 2005).

*The uncinate fasciculus* occupies the white matter of the rostral part of the temporal lobe, the limen insula, and the white matter of the orbital and medial frontal cortex. By connecting these temporal and prefrontal areas, the uncinate fasciculus may be a crucial component of the system that regulates emotional responses to auditory stimuli. It may also be involved in attaching emotional valence to visual information (Schmahmann and Pandya, 2005).

*The inferior longitudinal fasciculus* is in the white matter between the sagittal stratum medially, and the parietooccipital and temporal cortices laterally. It has a vertical limb in the parietal and occipital lobes, and a horizontal component contained within the temporal lobe. The ILF is the long association system of the ventral visual pathways in the occipitotemporal cortices (Schmahmann and Pandya, 2005).

*The fronto-occipital fasciculus* travels above the body and head of the caudate nucleus and the subcallosal fasciculus of Muratoff, lateral to the corpus callosum, and medial to the corona radiata. The FOF is the long association system of the dorsomedial aspects of the dorsal visual stream, and it appears to be an important component of the anatomical substrates involved in peripheral vision and the processing of visual spatial information (Schmahmann and Pandya, 2005).

*The cingulum bundle* nestles in the white matter of the cingulate gyrus. It is a composite fiber structure consisting of association fibers, striatal fibers, commissural fibers and subcortical fibers. The association fibers generally course through the periphery of the cingulum bundle, whereas the subcortical (commissural and subcortical) fibers tend to occupy a more ventral and central position (Talairach et al., 1973; Pandya et al., 1981; Mishkin, 1982; Squire and Zola-Morgan, 1991; Morris et al., 1999; Morecraft et al., 2001, Schmahmann and Pandya, 2005).

Corticostriatal fibers to the caudate nucleus, putamen and claustrum are conveyed mainly by the subcallosal fasciculus of Muratoff, and the external capsule.

The Muratoff bundle is a semilunar condensed fiber system situated immediately above the head and body of the caudate nucleus. It may be conceived of principally as a corticostriatal system from association and limbic areas. Additionally, it conveys fibers to the body of the caudate nucleus from the dorsal part of the motor cortex corresponding to the foot and trunk representations (Schmahmann and Pandya, 2005).

The external capsule appears to be predominantly a corticostriatal fiber system. It conveys fibers from the ventral and medial prefrontal cortex, ventral premotor cortex, precentral gyrus, the rostral superior temporal region, and the inferotemporal and preoccipital regions. Projections conveyed from primary motor and sensory cortices are directed exclusively to the putamen, whereas those from the supplementary motor area and association cortices terminate in the caudate nucleus as well (Schmahmann and Pandya, 2005).

The internal capsule is anatomically divided into the anterior and the posterior limbs. The anterior limb of the internal capsule conveys fibers from the prefrontal cortex including the rostral cingulated region principally to the thalamus, hypothalamus and basis pontis.

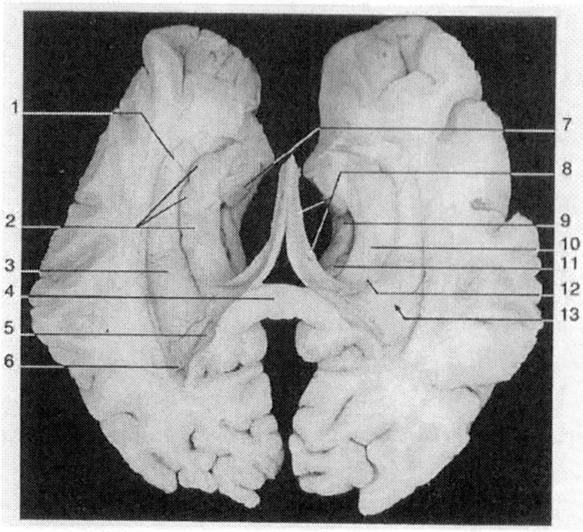


1. Temporal horn of the lateral ventricle
2. Hippocampus
3. Collateral parts
4. Splenium of corpus callosum
5. Calcar avis
6. Occipital horn of the lateral ventricle
7. Uncus of the hippocampus
8. Corpus of the fornix
9. Parahippocampal circumvolution
10. Hippocampus foot
11. Dentate circumvolution
12. Hippocampus phymbria
13. Lateral ventricle

Figure 7. An anatomical overview of the cingulum bundle.

The supplementary motor area fibers also lie in the anterior limb of the capsule.The posterior limb of the internal capsule conveys descending fibers from the premotor and motor cortices. Face, hand, arm, and leg fibers are arranged in a progressively caudal position. The posterior limb of the internal capsule also conveys descending fibers from the parietal, temporal and occipital lobes, and the caudal cingulated gyrus, and these are topographically arranged within the capsule, both in the rostral-caudal as well as in the superior-inferior dimensions (Schmahmann and Pandya, 2005).

Cortico-subcortical fibers enter the thalamus in specific locations determined by their site of origin. The afferent and efferent fibers are arrayed around the thalamus and are collectively termed the thalamic peduncles. The superior thalamic peduncle contains fibers from the caudal inferior parietal lobule, caudal cingulated gyrus and the dorsolateral prefrontal cortex. The lateral thalamic peduncle is located between the thalamus proper and the reticular nucleus, and conveys fibers from the superior and inferior parietal lobules, superior temporal gyrus and sulcus, cingulate gyrus, and motor cortices with topographic ordering. The anterior thalamic peduncle emerges from the anterior limb of the internal capsule, enters the most rostral part of the thalamus, and conveys fibers from all regions of the prefrontal cortex as well as from the supplementary motor area. The inferior thalamic peduncle emerges from the orbital frontal cortex and courses to the thalamus (Schmahmann and Pandya, 2005).

The ventral component curves upwards around the medial border of the globus pallidus, merges with the dorsal sector that lies medial to the globus pallidus rostrally and that penetrates the rostral thalamus at its medial and ventral aspects. It conveys fibers to thalamus from the caudal orbital cortex, the anterior and medial temporal lobe and the amygdale (Schmahmann and Pandya, 2005).

Fibers in the ventral subcortical bundle arise from the rostral parts of the superior temporal gyrus, superior temporal sulcus, inferotemporal area, and parahippocampal gyrus. They course medially over the tail of the caudate nucleus and the stria terminalis, destined primarily for the medial pulvinar (Schmahmann and Pandya, 2005).

In: Basal Ganglia and Thalamus
Editors: R. Moretti, P. Torre and R. M. Antonello
ISBN 978-1-60692-198-2


*Chapter 4*

# Basal Ganglia Region: Surgical and Anatomical Consideration

***Alessandro Melatini***
Consultant Neurosurgeon, Cattinara Hospital AOUTS , Trieste, Italy

Basal ganglia and immediate surroundings is a complicate anatomical region. Due to my professional duties, I often review anatomy before surgery in that area. In this chapter, I would like to delineate the three-dimensional structure and its intimate relationship, so that the reader will, hopefully, have a good mental image of the region. Understanding the relationship of the basal ganglia with the sub- arachnoidal space (and its mayor vessel), with the ventricular system and with the white matter is the key for surgery and anatomic-clinical correlations. You can find everywhere an atlas explaining the nuclei of basal ganglia; this is not the purpose of this chapter. We commonly use the term basal ganglia for the grey nuclei deeply located in the brain and so called because they are next to the cranial base. The exact definition of basal ganglia refer to caudatus putamen and pallidum. For the purpose of this chapter, I shall consider together, as an intimate complex, the thalamus, the basal ganglia, the internal capsule and the insula.

These regions form a unique block in every half of the brain which is an interesting concept from the surgical point of view. This 'block' is a central structure in each brain-hemisphere, and it is tightly related with the major part of the telencephalon and of the mesencephalon; thus, it seems a good start point to understand brain anatomy.

The block is a biconvex structure limited laterally by the shield of insula, medially by ependyma of lateral and third ventricle (a unique structure divided by the *tela chorioidea* in 2 compartments), behind by the surface of pulvinar thalami, frontally by the anterior surface of the insula and by the head of caudate nucleus, superiorly by the ependyma of the caudate nucleus and the thalamus, inferiorly by the ependyma of the temporal horn of the lateral ventricle. This ‘block’ is a continuous solution, inferiorly, with the mesencephalon structure, and superiorly with the telencephalon, through the corona radiata.

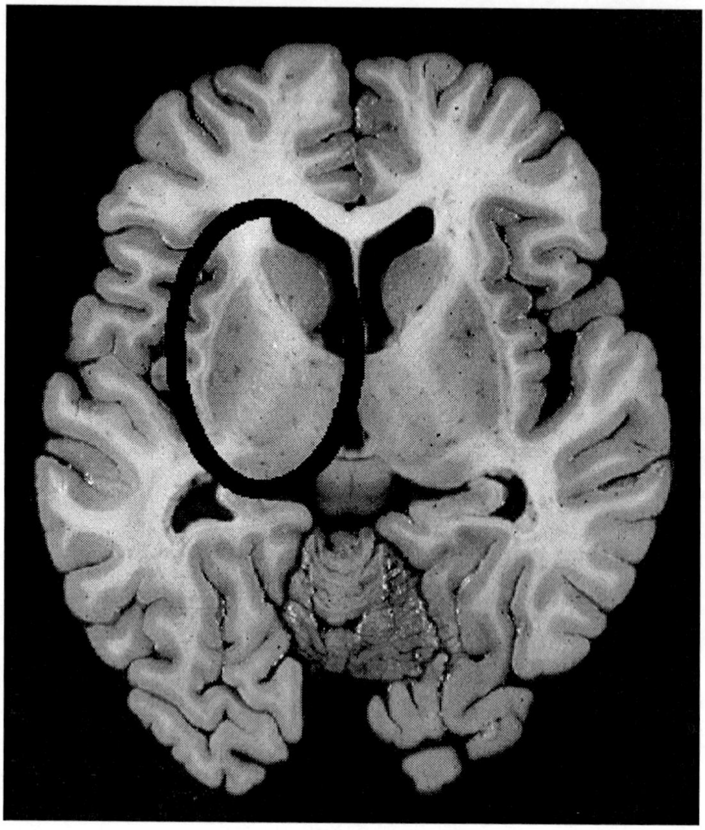

Figure 8. Lined in black, the block image (Modified by the Author).

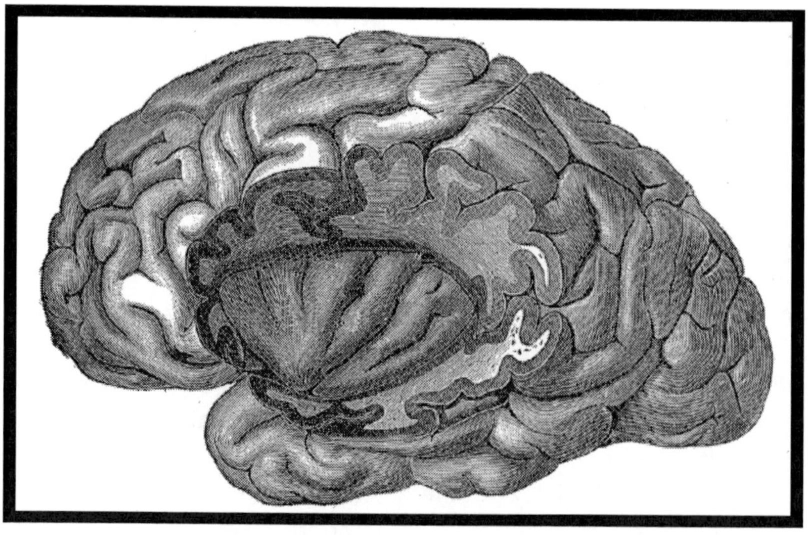

Figure 9. Lateral limits of the lock: removing the opercula of the frontal and the temporal lobes, the insula comes to view (modified by the author).

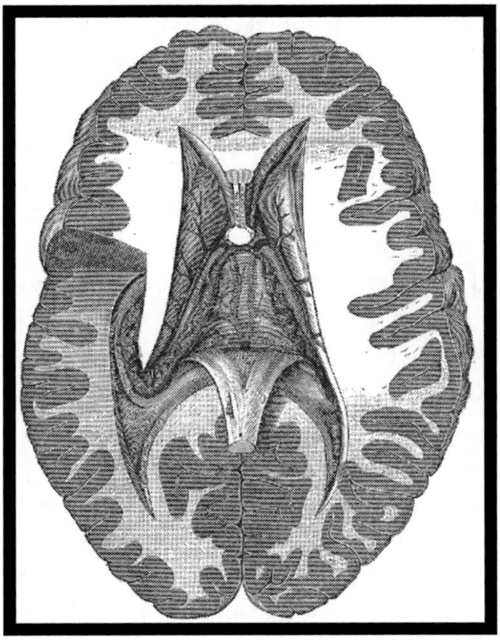

Figure10. The ventricular system, with its superior and medial limit; the tela chorioidea between the red choroids plexus (Modified by the author).

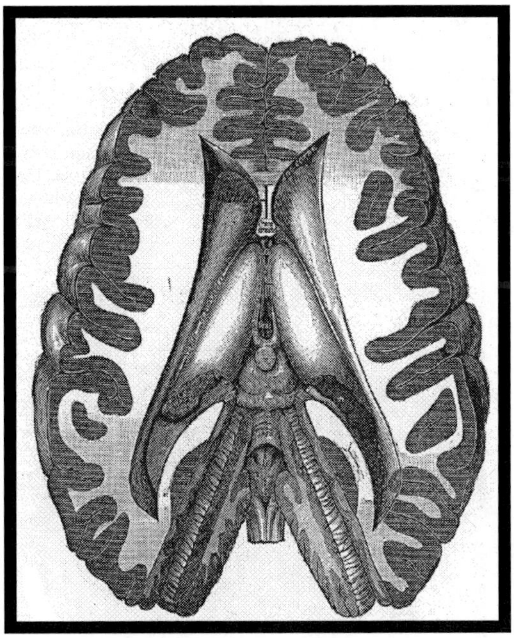

Figure 11. We have removed the tela chorioidea; thus, we have a panoramic view, at glance, of the lateral and of the third ventricle (limited by medial region of the thalamus) (Image modified by the author).

At this stage of the discussion, I shall make a gross division of the structure, to refine it later. The bigger part of deep gray matter is the medial region (thalamus and caudate nucleus); the cortex (of insula) is extremely lateral. For our debate, I shall consider the white matter (composed by the internal , the external and the extreme capsule) as a whole structure, and the putamen–pallidum complex, immersed in the white fibres, as a sort of second shield, protecting the internal capsule. In this context, the fibres from and to the thalamus and the caudate nuclei transit, connecting intimately the cortical and subcortical areas.

Up to the time, the first consideration is that the "block", which is commonly considerate as deeply submerged into the inner part of the brain, is not so real. Look at the structure formed by the insula, the basal ganglia and the thalamus as a unique nucleus. The major part of it surface (superior, medial and inferior) is lined by the ventricular ependyma, behind by the pia and the sub-arachnoidal space near the pulvinar, and laterally by the insula. All these surfaces are tightly related to cerebrospinal fluid, even though deeply located.

A smaller percentage of the block is a continuous "appendix" of the telencephalon and downwards through the mesencephalon. The ideal continuation is determined by the white substance (inferiorly) (via lemniscal, cortico-spinal, cortico-pontine, cortico-basal-cerebellar fibres), and superiorly (via cortico-spinal, thalamo-cortical, inter-lobar connections) through the mesencephalon. Building a simple scheme, the group insula basal ganglia thalamus is splitted by a fan positioned sagittally, and the fan represent all the fibres connecting the 'block' with the telencephalon. In relation to the fan, the thalamic and caudate nuclei are located medially. So, the entire ventricular system, which surrounds caudatus and thalamus, is located medially, in relation to the fan. Therefore, the optic radiations, a thalamo-cortical structure, is a lateral boundary of the ventricular trigone. The opposite happens with connecting white substance of the hemispheres. In fact, the fibres of the corpus callosum are located superiorly or medially, like the internal boundary of trigone of lateral ventricle.

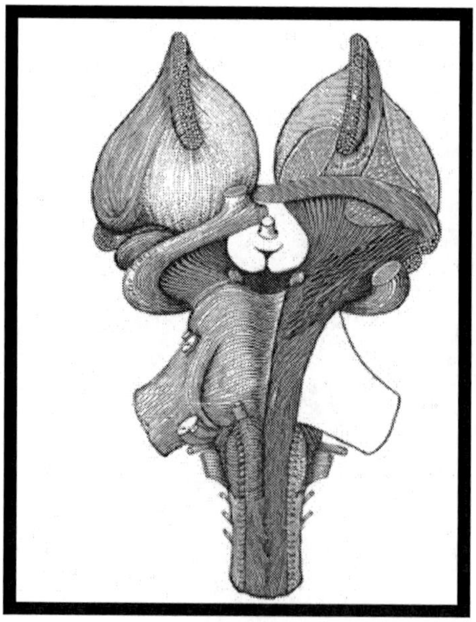

Figure12. An anterior view of the upper emerging fibres (image modified by the author).

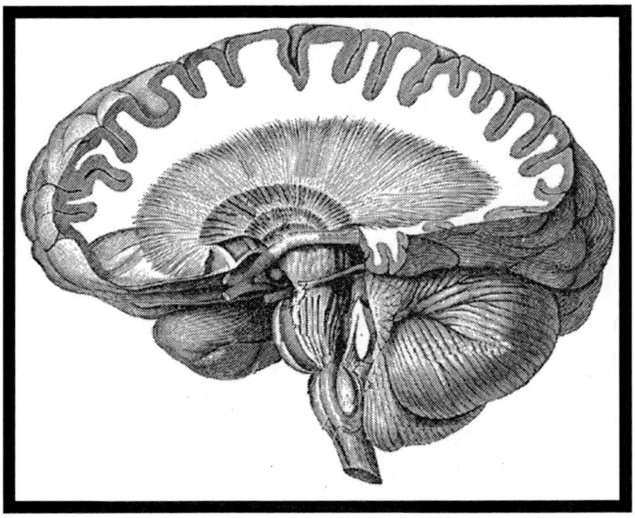

Figure13. Removing cortex of insula, the upper emerging fibres, forming corona radiate are visible (image modified by the author).

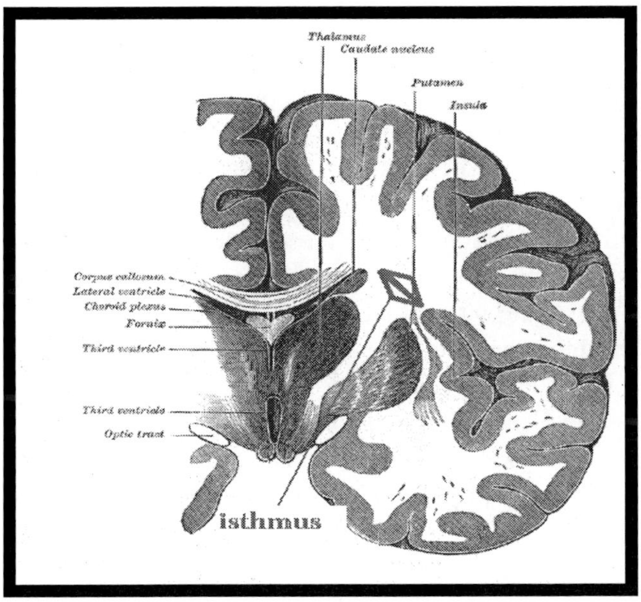


Figure 14. The rhomboid figure outline the isthmus, connecting white matter/basal region and telencephalon (image modified by the author).

The upper connection is limited medially by the lateral part of the ventricular body and laterally by the limitant of the insula. Through this slit, all the fibres of the internal capsule-corona radiate pass, as well as the fibres connecting frontal temporal and parieto-occipital lobes, like those passing through the external capsule and the capsula extrema.

This easy notion is the key for a safer trajectory in functional neurosurgery and an important limit in ablative surgery.

# Vascularisation of the "Block"

The insula, the deep nuclei, and internal capsule have different pattern of arterial supply.

The first has the same vascular supply of all the cortical surface, with the same characteristics of the leptomeningeal-pial network, with its high degree of anastomoses. The internal capsule, the thalamus and the basal ganglia are vascularized by a different number of perforating vessels, which possess anastomoses, but of smaller diameter (20-30 μm), in comparison with superficial anastomoses, whose calibre usually range from 20 to 90 μm.

The direct conclusion is that perforating vessels are unable to compensate acute decrease of blood flow, as it happens in a cardio-embolic event; on the contrary, they are capable to compensate a progressive flow reduction, which might occur in a progressive stenosis of the mayor arterial trunk (such as the internal carotid artery, the vertebral artery and their branches), or in peculiar clinical situation, such as Moya-Moya disease.

This reality is particularly useful during surgery, when an iatrogenic damage or the necessary coagulation of the superficial pial arteries have few (or fortunately enough) no consequence for the patient; just the opposite happens in case of a direct damage of the perforating arteries.

Let's now analyze the different blood supply of the central nuclei in a lateral-medial order. Most of the putamen is irrigated by perforators of the middle cerebral artery, except the rostro-ventro-medial part, which is nourished by Heubner's artery and anterior cerebral artery (A1). The Globus Pallidus is irrigated by perforators of internal carotid and anterior choroidal arteries in its medial part, while medial and anterior arteries supply respectively lateral and rostral portion. The anterior limb of the internal capsule is supplied by perforating branches of middle cerebral artery, Heubner's and some vessels from A1. The genu is irrigated by perforators of the middle cerebral artery, the anterior cerebral artery, the internal carotid, the anterior choroidal artery. The posterior limb receives blood from the perforating vessels of the middle cerebral artery, the internal carotid, the anterior choroidal, and partially by the posterior cerebral (thalamogeniculate) artery. The retrolenticular and sublenticular portions of internal capsule are supplied by the anterior choroidal artery, and only minimally by the thalamogeniculate perforators (posterior cerebral artery).

The head and the body of the caudate nucleus are nourished by perforators from Heubner's artery, A1, by the middle cerebral artery; the proximal part of the tail of the caudate nucleus is irrigated by the branches of the lateral posterior choroidal artery, while the distal part is irrigated by the distal perforators of the anterior choroidal artery. The thalamus has the most complex vascularization; its afferents derived from the anterior, the posterior circle and from the posterior communicating artery (PCo). Anterolateral thalamus is supplied by the premammillary artery, a branch of PCo. The posterolateral thalamus receives blood supply from the thalamogeniculate perforators, arising from P2 segment of the posterior cerebral artery. The lateral thalamus is nourished by the anterior choroidal artery. The medial part of the nucleus receives from thalamo-perforating branches of P1 portion of the posterior cerebral artery. The dorsal thalamus is supplied by the medial, the lateral choroidal arteries, and the cingulothalamic artery. The pulvinar receives blood supplementation from the anterior, the posterior the medial and the lateral choroidal arteries. The subthalamic nucleus receives from the anterior choroidal and the posterior communicating arteries.

# Pathology and Clinic of Vascular Lesion: The Neurosurgery Perspective

General diseases, like heart disease, atherosclerosis, diabetes mellitus , polycytemia might cause micro-embolic infarction localized in the territory of perforating arteries. Cerebrovascular disease comprises different processes, such as atherosclerosis and consequent ischemia, aneurysms, subarachnoid haemorrhage , subintimal haemorrhage, dissection and vasculitis. All of them interest in different ways the perforating arteries. Atheromas may spread embolic fragment in patent vessel and or perforating branches. Primitive subarachnoid hemorrhage causes indirectly and directly a conspicuous spasm of perforating arteries. Aneurysms compress, distort, or stretch the perforating vessel. Sub-intimal haemorrhage and dissection usually involve the orifice of one or more perforators. Atheromas, fibrinoid necrosis, lypohyalinosis affect the wall of the perforating vessels, causing a rupture and consequent intraparenchimal bleeding.

Vascular malformation, due to their nature, lead to rupture. The regions mainly interested by aneurisms are the carotid bifurcation, and middle cerebral artery (MCA) bifurcation; the artero-venous malformation mainly interest the deep brain structures, and almost ubiquitary are the cavernous angiomas.

Surgery can damage the same arteries.

As a direct consequence of the vascular supply of the block, the perforating vessels irrigate different structures (i.e. perforators of MCA supply the caudate nucleus, the putamen, the globus pallidus and portion of internal capsule); hence, the occlusion of these arteries result in large infarct of basal ganglia, and surrounding structure. On the other hand, each part of basal ganglia is supplied from several sources. And mainly due to this second point, the occlusion of a perforating vessel can cause only a partially limited ischemia.

The most typical aspect of this pathological event is the lacunar infarct of the posterior limb of the internal capsule, which causes a controlateral hemiparesis. In addition, the extension of brain ischemia, through the cortico-ponto-cerebellar fibres, causes the incoordination of arm and leg, configuring the so-called ‘capsular ataxic emiparesis’. Dysarthria and clumsy hand syndrome is frequently caused by infarct of the dorsal genu of the capsule. When the motor disturbances is associated with an acute evidence of confusion and behaviour disorders, one might expect the clinical finding of two synchronous infarcts, of the anterior and of the posterior limb of the internal capsule.

Regarding the basal ganglia, the ischemic lesion of different nuclei may lead to the same clinical picture. Typically, lesions of the subthalamus cause hemiballism, but the same motor disturbance is associated with lesion of the neostriatum. Monolateral lesion of the globus pallidus can even be asymptomatic, but bilateral lesions lead to global hypokinesia. Vascular lesion of the neostriatum is associated with dyskinesia, chorea, hemiballism, with a general reduction of the spontaneous behaviour, a sensory neglect, dysphasia and other cognitive disorders. The clinical picture of thalamic vascular alterations reflects the anatomical division, which delineate four vascular zones, above mentioned. An anterolateral infarct (territory of the pre-mammilary artery, a branch of PCo artery) causes slight hemiparesis, dysphasia (if left), impairment of intellect, memory, visual perception and constructional praxia. A posterolateral lesion leads to hemianopsia, hemi-dysesthesia, ataxia, visuospatial neglect, disorder of attention, hemiparesis, and sometimes, thalamic pain. Medial infarct determines an impairment of intellect, memory, visual perception, constructional praxia, and orientation in time. A lateral lesion causes evident disturbances of visuospatial perception and memory, as well as anosognosia, hemianopsia, hemiparesis. Dorsal infarcts are very rare due to the high number of anastomotic network, which characterized the lateral and posterior choroidal arteries.

## Deep Nuclei and Neoplasm Behaviour

During surgery of intrinsic CNS tumours, like glial neoplasms, the most common fear of the surgeon is to invade, with resection, vital structures such as cortico-spinal tract, or basal ganglia, and their connecting fibres (i.e. the internal capsule). The concept of infiltration in CNS is very different from the other body neoplasms. Especially with a concern to high grade glioma, the long white pathways constitute a real barrier for the expansion of the tumour. The high-grade tumour, by itself, has a sharp limit determined by the surrounding white matter, and the surgeon should not rescinds over the virtual limit (change of the tissue's colour and density) viewed through the field determined by the operating microscope. It is a common experience in neuro-surgery to find a sharp change, from the brown-yellow colour of the tumour tissue, and the normal (or slightly oedematous) white matter. In the same way, as experienced for long pathways, the diffusion of the tumour from sub-pallial and pallial areas to the central nuclear area, and vice versa, does not occur commonly.

Primary tumours of central nuclei are rare. They develop from glial cells, supporting white matter fibres that run among and through the gray matter of these nuclei. Such lesions arise mainly from the head of the caudate nucleus, the thalamus and the hypothalamus. Almost exceptional are the tumours primarily located in the putamen, pallidum and claustrum. As for the caudate nucleus, neoplasms tend to localize in the head and body, but the tail is usually spared. Usually these tumours expands centrifugally, in all the directions. Growing, they distort and displace all the surrounding structures, like the internal capsule and the adjacent nuclei, though they respect their anatomical border. In the same way, they distort the ventricular wall; on the contrary, very rare is the disruption of the ependyma and of the glial membrane, with a consequent invasion and dissemination of the ventricular system. Infiltration of the long projection, associative, and commissural fibres, as mentioned above, does not occur, a part from the late progression phases of the tumour-life.

This last observation is the ground for the clinical-operative decision to avoid surgery, because of the neoplasm' behaviour, in these circumstances, do not respect the above mentioned "rules", and therefore, the lesion's neoplastic progression is so advanced that surviving rate is very poor.

## Surgical Consideration

What previously said is very important for the neurosurgeon, performing different types of surgical procedure.

Stereotactic surgery, through needle or electrode insertion, is essentially a "blind procedure". The surgeon has an indirect vision of the region to cross with the needle through stereo-radiological images; through them, he could simulate the trajectory, supposed to be safer. Nevertheless, he has not the direct control of the tip of the needle, and nothing can be done to avoid, to control and to repair the damage, such as a vessel rupture, with consequent haemorrhage. The visual control is possible only in cortical surface, and it is virtually lost few centimetre below.

If an incorrect, or erroneous procedure has been made, a bad plan occur. This way, it happens that the needle crosses the deep part of a sulcus and its subarachnoid vascular structure: thus, bleeding could occur. In this case, and in case of deeper accidents, we can only observe indirect signs of damage, such as brain swelling. In so far, the safer way still is to pass through the brain parenchyma, in order to avoid subarachnoid space and its vasculature, easier to damage. In the same way, the cross of the ventricular system could damage the deep strio-thalamic vein, or the choroids plexus vasculature, or the subependymal vasculature, whose wall is particularly fragile and a very small trauma can lead to its rupture. Anyway the latter case is less frequent in complication as expected .

On the other hand, the anxiety of the surgeon (or at least my one) for multiple needle blind insertion in functional stereotaxis seems to be unjustified, reviewing clinical experiences all over the world. Observing cautiously anatomy, the surgeon can reduce the risk of haemorrhage at an acceptable level, mainly inferior to 3 % for functional stereotaxis, sometime higher for tumour biopsy, depending from the neovascularization of the neoplastic lesion.

In ablative microsurgical procedures, the surgery is quite 'easy'. As mentioned before, paradoxically, deep nuclei are not so deep. As previously stated, the most of their surface is in contact with real of virtual spaces, like ventricular system, and subarachnoid space. It means that the trajectory to reach the lesion is predominantly extra cerebral, or in some case purely via subarachnoid space. For the latter, lets make an example concerning a glial neoplasm of posterior inferior thalamus. Posterior-inferior thalamus surface is localized in part in direct contact with the subarachnoid space, and in lateral part near the ventricular system. The optimized surgical access will be in a space between the tentorium and the superior surface of cerebellum, in order to reach the thalamic surface, laterally to the Rhosental vein and medially to the choroidal fissure. It might results as a long way, but it spares every cerebral structure, as far as it results to be a pure subarachnoid access. In comparison, a trans ventricular way, crossing 2-5 centimetres of the white substance, would invariably open a trajectory-injury.

Therefore, the key for "minimal invasive" neurosurgery is a good knowledge of anatomy, and, in a second line, the skilful aid of technology .

In: Basal Ganglia and Thalamus
Editors: R. Moretti, P. Torre and R. M. Antonello
ISBN 978-1-60692-198-2


*Chapter 5*

# Physiology of Basal Ganglia and Thalamus Circuits

***Rita Moretti, Paola Torre, Rodolfo M. Antonello, and Antonio Bava****
* Dip. Fisiologia e Patologia, Università di Trieste, Italy

In the present and most restrictive use, the basal ganglia is generally composed by three groups of brain structures called the "striatum," "pallidum," and "substantia nigra," and an additional structure termed the "subthalamic nucleus" (STN). The striatum can be separated into two general components, the dorsal striatum, which consists of the caudate and putamen, and the ventral striatum, which consists of the nucleus accumbens, septum, and olfactory tubercle. The nucleus accumbens, in turn, is subdivided into a lateral care region and medial region. In a similar manner, the pallidum can also be divided into multiple divisions, including a lateral or external segment of the globus pallidus (GPe), a medial or internal segment of the globus pallidus (GPi), and a portion that lies ventral and anterior to the anterior commissure, designated the ventral pallidum (VP). The GPi is further divided anatomically by a distinct fiber bundle into a lateral outer portion and a medial inner portion. Finally, the substantia nigra is also composed of more than one component, a cell group rich in neuromelanin called the pars compacta (SNpc), which is responsible for the black appearance of the nucleus in gross specimens, and an unpigmented cell group known as the pars reticulata (SNpr). Further subdivision of the SNpr is also possible, due to the presence of a distinct subpopulation of cells located dorsolaterally in the SNpr, and is referred to as the pars lateralis (Middleton, 2003).

According to the simplest classification scheme, the nuclei within the basal ganglia can be subdivided into sets of "input," "output," and "intermediate" structures. Input structures include the caudate, putamen, and nucleus accumbens. Collectively, these input structures receive direct projections from nearly the entire cerebral cortex. These input structures then project to intermediate structures as well as output structures.

Intermediate structures within the basal ganglia include the STN, GPe, and SNpc. An additional structure, the pedunculopontine nucleus (PPN), is also occasionally viewed as an intermediate component of the basal ganglia due to its dense interconnectivity with other basal ganglia structures. In general, intermediate structures project most heavily to other basal ganglia nuclei, including the output structures as well as other intermediate structures (Middleton, 2003).

The three principal output structures of the basal ganglia include the GPi, SNpr, and VP. For the most part, these output structures send their efferent projections to different subdivisions of the ventroanterior-ventrolateral (VAIVL), mediodorsal (M,D), and intralaminar (IL) groups ofthalamic nuclei (particularly the centrum medianum and parafasicularis intralaminar nuclei *(*CM/PF*)* (Middleton, 2003). The VA/VL and MD nuclei of the thalamus, in turn, project largely back upon the cerebral cortex. Thus, one of the major features of basal ganglia anatomy is their participation in what has become known as cortical-basal ganglia-thalamo-cortical circuits (hereafter referred to as simply cortical-basal ganglia circuits).

Notably, however, some of the neurons in VA/VL and MD, and many of the neurons in CM/PF, send projections to the striatum, and thus form a recurrent thalamic feedback loop with the basal ganglia.

Most of the cortical input to the basal ganglia is derived tram layer V pyramidal cells that use glutamate (GLU) as a neurotransmitter. These cortical projection neurons can innervate one or more of several types of basal ganglia circuits (Middleton, 2003).

Many of these same cortical areas also contain cells that project directly to the STN and, thus, constitute a subthalamic path (Middleton, 2003). Notably, while the striatal inputs from motor and higher association areas in the cerebral cortex target neurons in the striatal matrix, which contain high levels of acetylcholinesterase, the inputs from many orbitofrontal and temporal limbic regions of the cortex target regions of the striatum called patches, or striosomes, which express low levels of acetylcholinesterase (Middleton, 2003). Neurons in the striosomes do not participate in the same types of circuitry as neurons in the striatal matrix. While medium spiny neurons in the matrix project largely to the gamma-amino butyric acid (GABA)-ergic cells of the pallidum and nigra (GPe, GPi, or SNpr), the medium spiny cells in the striosomes project largely to dopamine synthesizing cells in the SNpc (Middleton, 2003).

Thus, at least at the input stage of processing, a fourth type of circuit, the cortico-striosomal path can be envisioned.

SNpc cells project densely back to widespread regions of the striatum (as well as the STN and pallidum) where they maintain high levels of extracellular and synaptic dopamine pools. The net effect of the dopamine on striatal cells depends largely an the types of receptors present on the target cells.

For example, striatal cells in the direct path (that project to the GPi or SNpr) express high levels of D1 receptors, as well as substance P and dynorphin (Middleton, 2003). Activation of D1 receptors on these cells results in increased cellular activity and responsiveness through G(s)-mediated second messenger systems.

In contrast, striatal cells in the indirect path (that project to the GPe) express high levels of D2 receptors as well as enkephalin (ENK) (Middleton, 2003; Alheid et al., 1990). Activation of D2 receptors on these cells results in decreased cellular responsiveness and activity through G-mediated second messenger mechanisms.

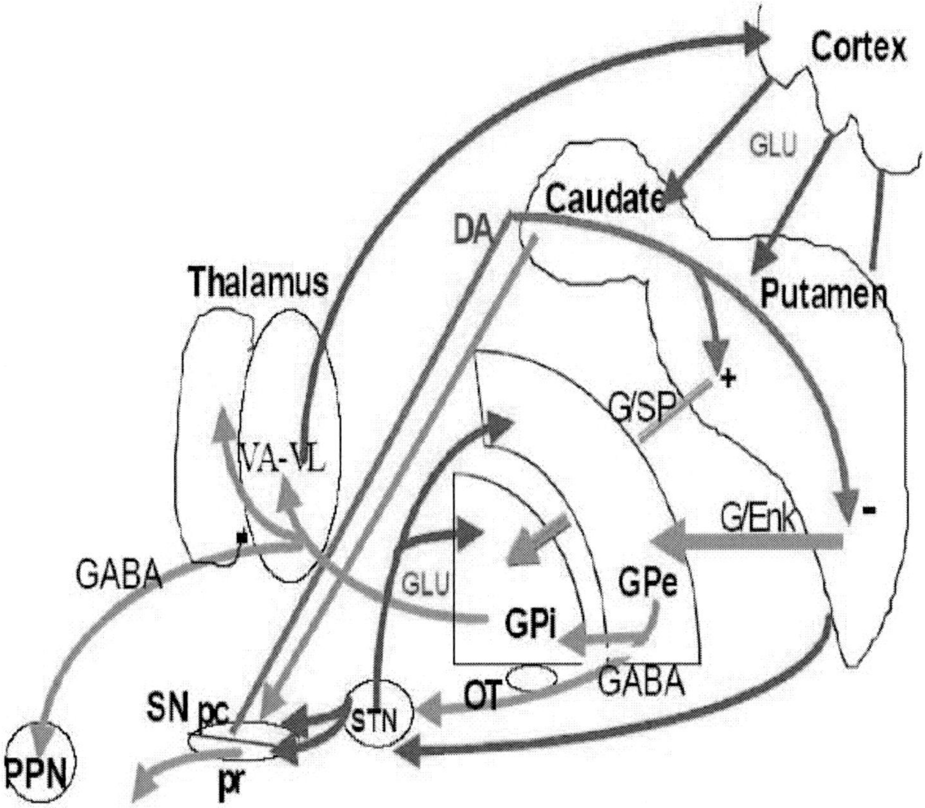


Figure 15. The cortical output is excitatory to striatum and STN; the striatal, GPe and GPi outputs are inhibitory; the STN and thalamic output is excitatory.The nigrostriatal pathway is dopaminergic.

Despite the differences in the effects of dopamine an the direct and indirect pathways, the medium spiny cells in both of these circuits utilize GABA to inhibit their targets. Because the GPe cells that receive D2-mediated input project largely to the GPi, then the net effect of activation of the direct and indirect pathways in the presence of normal levels of dopamine is the same in terms of information flow; that is, cortical excitation of the basal ganglia input structures leads to less inhibition of the thalamus and increased activity of these circuits (Middleton, 2003).

Importantly, however, cells in the GPe also send inhibitory inputs to the STN, which in turn exert a powerful excitatory influence an the GPi and SNpr. Thus, D2-mediated inhibition of the GPe can promote activity of the STN and allow the basal ganglia to increase inhibition of the thalamus. This effect would be further promoted by the action of direct cortical excitatory inputs to cells in the STN.

In the striatum, one important class of interneuron is the large or giant cholinergic interneuron. These neurons receive most of their input from thalamic cells in CM/PF and other striatal cell types and a small input from the cortex. In turn, giant cholinergic cells have widespread projections to other striatal cells, largely within the matrix (Middleton, 2003). Giant cholinergic cells also express high levels of acetylcholinesterase activity, as well as D1, D2, and selected muscarinic and glutamatergic receptors (Kawaguchi et al., 1995; Middleton, 2003).

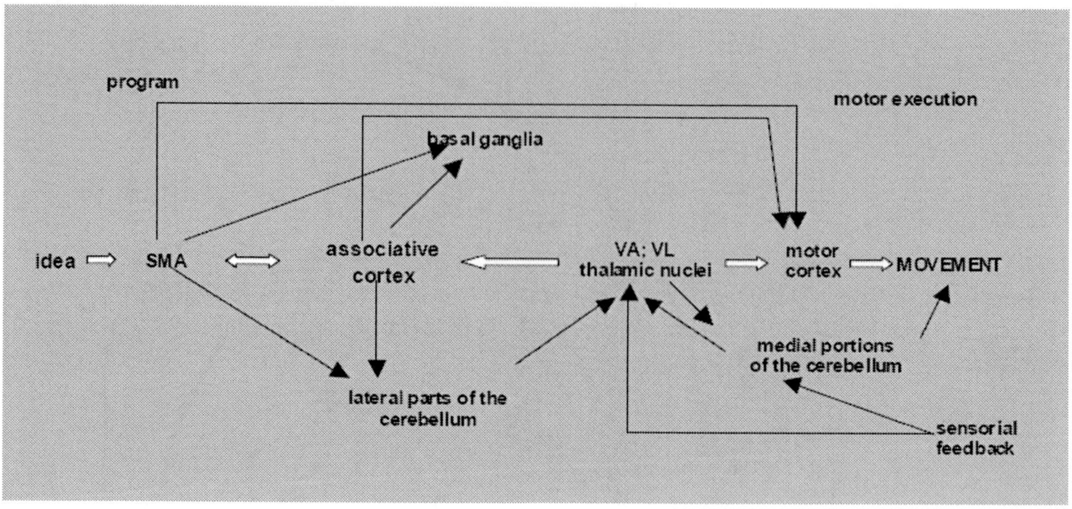


Figure 16. The on-line operative control exerted by the basal ganglia.

One of the more recent controversies concerned the issue of how many cortical areas participated in basal ganglia circuits, and how the information from different cortical areas was processed within these circuits. Historically, at the input level of processing, the projections from the cerebral cortex to the basal ganglia were recognized as a highly organized and topographic projection system (Alexander et al., 1986; Midlleton, 2003). For example, anterior cortical areas were seen to project to anterior striatal regions, and posterior cortical areas were observed projecting to posterior striatal regions. Likewise, ventromedial regions of cortex projected to ventromedial striatal regions, whereas dorsolateral areas of cortex projected to dorsolateral striatal regions (Middleton, 2003).

However, beyond these features, most theories of basal ganglia function held that there was a great deal of convergence both within the striatum and in the projections from the striatum to the intermediate and output levels of processing.

In other words, the basal ganglia were thought to have the ability to integrate sensory, limbic, and cognitive information with the commands for movement.

There were two main reasons far the widespread acceptance of this viewpoint. The first reason was that the symptoms of basal ganglia dysfunction most often include characteristic motor abnormalities.

Any cognitive or emotional disturbances also seen in patients with basal ganglia disease were often attributed to other causes and ruled secondary. The second major reason that the basal ganglia were viewed solely as a motor-related structure was the lack of substantive evidence far basal ganglia projections back to non-motor regions of the cerebral cortex.

Alexander, DeLong, and Strick (1986; Midlleton, 2003) reviewed the results of numerous anatomical studies and suggested that rather than serving as a tunnel far information from widespread cortical areas, the basal ganglia actually participated in multiple parallel segregated circuits with different regions of the frontal lobe. These regions included cortical areas concerned with skeleton-motor and oculomotor control and three regions of the prefrontal cortex involved in cognitive and limbic functions. Recent experimental evidence has supported and expanded on this proposal. More recently (Middleton and Strick, 2000), it was suggested that the original five circuit scheme, proposed by Alexander and colleagues (1986), should be broadened to include seven general categories of circuits: skeletomotor, oculomotor, dorsolateral prefrontal, lateral orbitofrontal, medial orbitofrontal, cingulate, and inferotemporal-posterior parietal.

Within each of these categories, anatomical evidence supports the existence of multiple (between two and seven) parallel cortical-basal ganglia circuits.

By virtue of their reciprocal connections with posterior cortical areas involved in visual and spatial perception, the basal ganglia would have the potential ability to influence an even broader range of behaviours than those proposed in the model by Alexander and colleagues (1986). In that model, the basal ganglia inputs tram posterior cortical areas were integrated in basal ganglia circuits and ultimately influenced only regions of the frontal lobe.

Finally, although there is considerable evidence that much of the basal ganglia circuitry comprising the direct, indirect, and subthalamic pathways is topographically organized at multiple levels, there is much less evidence to suggest that the striosomal path is similarly organized.

Indeed, the recurrent projections from the SNpc to widespread regions of the striatum may represent the means far the basal ganglia to integrate information about primary reward and behavioural state received from limbic and sensory cortical areas with the modality-specific information being processed in the cortical-basal ganglia circuits.

As far as motor control is concerned, a well conducted study defined the basal ganglia role in it.

The results from three groups of PET studies of regional CBF show a double dissociation between selection of movements, which induces differential effects in the basal ganglia but not the cerebellum, and sensory information processing, which involves the cerebellum but not the basal ganglia (Jeuptner and Weiler, 1998). The first set of studies concerned motor learning of a sequence of finger movements; there was a shift of activation in the anterior–posterior direction of the basal ganglia which paralleled changes in the motor areas of the frontal cortex. During new learning, the dorsolateral prefrontal cortex and striatum (caudate nucleus and anterior putamen) were activated. When subjects had to select movements, the premotor cortex and mid-putamen were activated. With automatic (overlearned) movements, the sensorimotor cortex and posterior putamen were activated. When subjects paid attention to overlearned actions, activation shifted back to the dorsolateral prefrontal cortex and striatum. The cerebellum was not activated when subjects made new decisions, attended to their actions or selected movements(Jeuptner and Weiler, 1998).

These results demonstrate components of basal ganglia– (thalamo)-cortical loops in humans. According to earlier studies in animals the Authors proposed that the basal ganglia may be concerned with selecting movements or the selection of appropriate muscles to perform a movement selected by cortical areas (e.g. premotor cortex) (Jeuptner and Weiler, 1998).

Secondly, a visuomotor co-ordination task was examined. In the absence of visual control over arm movements, subjects were required to use a computer mouse to either generate new lines or to re-trace lines on a computer screen (Jeuptner and Weiler, 1998).

The neocerebellum (hemispheres of the posterior lobe, cerebellar nuclei and cerebellar vermis), not the basal ganglia, was more engaged when lines were re-traced (compared with new line generation). Animal experiments have shown that error detection (deviation from given lines) and correction occurs during line re-tracing but not line generation. These data suggest that the neocerebellum (not the basal ganglia) is involved in monitoring and optimizing movements using sensory (proprioceptive) feedback (Jeuptner and Weiler, 1998).

Thirdly, the relative contribution of sensory information processing to the signal during active/passive execution of a motor task (flexion and extension of the elbow) was examined; it was found that 80–90% of the neocerebellar signal could be attributed to sensory information processing. The basal ganglia were not involved in sensory information processing. They may be concerned with movement/muscle selection (efferent motor component); the neocerebellum may be concerned with monitoring the outcome (afferent sensory component) and optimizing movements using sensory (feedback) information (Jeuptner and Weiler, 1998).

In: Basal Ganglia and Thalamus
Editors: R. Moretti, P. Torre and R. M. Antonello
ISBN 978-1-60692-198-2


Chapter 6

# How are Basal Ganglia and Thalamus Related to Cognition and Behaviour?

***Rita Moretti, Paola Torre, Rodolfo M. Antonello, and Antonio Bava****
* Dip. Fisiologia e Patologia, Università di Trieste, Italy

The anatomical substrate exists for the basal ganglia to process large amounts of information related to skeletomotor, oculomotor, cognitive, limbic, and sensory functions.

There are different data, either from a clinical either from an experimental perspective, that suggest that basal ganglia are tightly related to cognition and behaviour, not through a *per se* function, rather via their dense and widespread projections (efferent and afferent) to different cortical and limbic areas.

Areas 9, 46, and 12 are now known to participate in basal ganglia circuits. Areas 9 and 46 project to adjacent regions of the dorsal and lateral caudate, which in turn innervate adjacent regions of the rostral and dorsal GPi and SNpr. These regions of the GPi and SNpr are now known to project back upon areas 9 and 46 via neurons in VANL and MD (Middleton and Strick, 2000; Middleton, 2003), thus completing the parallel circuit with these cortical areas. Area 12, in contrast, projects more ventrally in the caudate and appears to receive its input largely from the SNpr, but not GPi (Middleton and Strick, 2000; Middleton, 2003).

The physiological properties of the dorsal and lateral prefrontal cortex suggested that areas 9, 46, and 12 each appear to be involved in at least four different types of cognitive functions. First, many neurons in these areas display changes in activity related to the performance of delayed-response tasks. These tasks require sensory cues to be stored for a brief period of time and then used to generate a specific response (Rosenkilde, 1979;Watanabe, 1986; Goldman-Rakic, 1987; Goldman-Rakic, 1990; Passingham, 1993; Fuster, 1997; Middleton, 2003).

During such tasks, selective changes in neuronal activity con be seen during the presentation of cues (cue-related activity), the delay period following cue presentation (delay-related activity), and the response preparation and execution periods (response-related activity) (Rosenkilde, 1979; Goldman-Rakic, 1987; Goldman-Rakic, 1990; Passingham,

1993; Fuster, 1997; Middleton, 2003). Depending on the area of the prefrontal cortex in which one records or studies, there are additional task parameters that will determine the relative proportion of neurons involved in the performance of these delayed response tasks (Middleton, 2003).

For example, within areas 9 and 46, many neurons have cue- and delay-related activity that is tuned to the spatial position of stimuli (spatially-tuned activity) or the order in which the stimuli are present (sequentially-tuned activity) (Middleton, 2003). In contrast, within area 12, neurons appear to be much less involved in spatial and sequential information and more involved in remembering the identity of particular objects (object-tuned activity) (Middleton, 2003).

Throughout areas 9, 46, and 12, there are also many neurons that appear to be more involved in the learning and remembering of specific rules or associations used in the performance of conditional response (forced choice) tasks (rule-related activity) (Middleton, 2003). Finally, a relatively small but consistent proportion of neurons in areas 9, 46, and 12 also display changes in activity that coincide with the expectation or anticipation of primary reinforcement or food rewards (reward-related activity) (Yamatani et al., 1990; Middleton, 2003).

Lesion studies support some of the physiological observations regarding areas 9, 46, and 12. Bilateral damage to area 9 in monkeys produces severe and long-lasting impairments in the performance of tasks that require animals to monitor the order and identity of different objects (but not their spatial position), particularly when the sequence of cues to be remembered is self-ordered (Petrides, 1995; Middleton, 2003). These deficits are similar to those found in human patients with dorsal prefrontal lesions (Petrides and Milnes, 1982; Middleton, 2003).

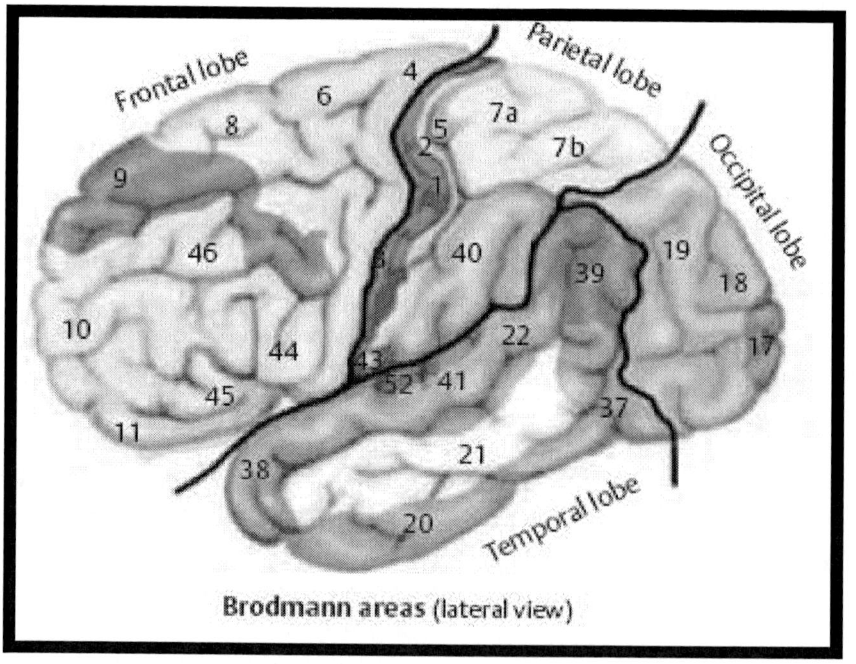


Figure 17. A synopsis of Broadmann cortical areas.

In addition, patients with prefrontal lesions have been shown to be deficient in tasks that require the categorization and sorting of different stimuli (Milner, 1063; Middleton, 2003), or the planning and monitoring of script events or sequential actions (Pascual-Leone et al., 1995, Middleton, 2003). These types of tasks ore thought to impose significant demands on planning and short-term (or working) memory far objects or sequential actions. Experimental lesions of area 46 result in an inability of animals to perform spatial delayed-response tasks, spatial delayed-alternation tasks, and go/no-go tasks (Sirigu et al., 1995; Middleton, 2003).

Finally, lesions of area 12 in monkeys have been shown to produce what has been referred to as an inability of animals to make switches in behavioural set. This deficit leads to perseverative responses on delayed-response tasks, in which the identity of different objects or colours must be remembered (object matching, object reversal, or object alternation tasks), or inappropriate responses in auditory-cued go/no-go tasks (Mishkin and Manning, 1995; Middleton, 2003).

In addition, there is also evidence that large lesions of area 12 produce deficits in the learning and performance of visual discrimination tasks that do not involve delayed-responses (Rushworth et al., 1997; Middleton, 2003).

Functional imaging studies in humans support many of the physiological and behavioural findings.

Tasks that require difficult planning, the monitoring of sequences of hand movements, or the learning of new movement sequences (Sakai et al., 1998; Middleton, 2003) produce sites of peak activations in lateral portions of area 9, as do tasks that employ verbal working memory (including word generation tasks) (Paulesu et al., 1993; Middleton, 2003). Similar to area 9, area 46 is particularly active during spatial working memory tasks, difficult planning tasks, go/no-go tasks, some non- spatial object and verbal working memory tasks, and verb generation tasks (Grasby et al., 1993; Middleton, 2003). Area 46 has also been shown to be active during the generation and monitoring of multiple movement sequences involving the hand or fingers (Jeuptner et al., 1997; A-B; Middleton, 2003). Many of these studies have concluded that the prefrontal activations in areas 9 and 46 are related to the elaboration or learning of novel sequences, since these activations are most apparent when comparing task performance during the learning phase with that seen after the task is well learned (Middleton, 2003). The human equivalent of area 12 (area 47) has been shown to be active during tasks that employ verbal working memory, including word generation tasks (Paulesu et al, 1993; Fiez et al., 1996; Middleton, 2003), but not during spatial working memory tasks.

Directly, per se, basal ganglia seem to exert some properties among cognition and behaviour.

In a study that examined the effect of task difficulty on caudate nucleus properties in humans, Abdullaev and colleagues (1998) recorded from human subjects during reading, naming, recognition memory, categorization, and lexical decision making tasks. They found that, during visual processing of words, caudate cells exhibited excitatory responses related to both semantic and phonological-articulatory encoding and that the delay-related firing of cells was increased whenever semantic processing was required (Abdulaev et al., 1998; Middleton, 2003). Comparing the properties of striatal cells with those seen in the prefrontal (Broca's) area and in the temperal and parietal lobe areas, the authors found that the caudate cells had properties that were "strikingly similar" to those in Broca's area. Since portions of the prefrontal cortex are often activated during the same types of verbal tasks that activate Broca' s area, it is possible that basal ganglia inputs from these areas also contributed to the findings of this study (Middleton, 2003).

Striatal cell activity during sequential task performance has also been examined. Kermadi and Joseph (82) examined the properties of striatal cells during a task that required monkeys to engage in a sequential working memory test. They found that many of the cells with task-related activity had visual-related responses that varied according to the order of a given target, and moreover, that many of the caudate cells seemed to anticipate the fixation of specific targets. Such findings are similar to some of those reported in studies of areas 9 and 46 (Middleton, 2003).

Shuvaev and Shefer (1995) in another conditional response task in monkeys found that the task-related striatal cells they recorded could be separated into two distinct groups. While one group of cells was clearly tuned to response execution, another group of cells appeared to be more involved in the instructional decision-making process (Middleton, 2003).

Nishino and colleagues (1981) found that a small proportion of neurons in the head of the caudate nucleus of monkeys responded specifically to the sight of food or food rewards. Furthermore, the results from several recent studies strongly suggest that such reward-related properties can directly affect the cognitive properties of striatal cells.

In a similar finding, Hassani and colleagues (2001) observed that during a spatial-delayed response task, in which different picture cues were associated with specific rewards, many neurons in the anterior caudate, putamen, and ventral striatum of monkeys displayed different levels of task-related activity depending on the type of reward being offered and whether it was a preferred or non preferred reward of that animal (HAssani et al., 2001; Middleton, 2003).

These regions of the GPi and *SNpr* largely coincide with the regions that have been shown to project to skeletomotor or oculomotor regions of cortex (Middleton and Strick, 2000). In contrast, large portions of the GPi and *SNpr* contain neurons whose activity is not modulated by simple skeletomotor or oculomotor tasks. Many of these regions fall within the regions that innervate areas of prefrontal cortex involved in cognitive processing (Middleton and Strick, 2000; Middleton, 2003). Hikosaka and Wurtz (1983) and Hikosaka and colleagues (1993) recorded the activity of neurons in the *SNpr* of monkeys trained to perform an oculomotor spatial-delayed response task. A considerable number of *SNpr* neurons were found whose activity was modulated during the delay period of the task. Many of these delay-related cells appeared to be separate from *SNpr* cells, which were purely saccade-related, but not delay-related, and a number of them were spatially tuned. Neuronal properties similar to these have been found in studies of area 46 (Middleton, 2003). Thus, nigral outputs to prefrontal cortex could be directly involved in tasks that involve spatial working memory and the planning of eye movements (Middleton, 2003).

Thus, spatial and sequential aspects of delayed task performance were found within basal ganglia circuits that subserve prefrontal areas known to be involved in the same type of tasks (Middleton, 2003).

In one of the earliest studies of pallidal physiology in awake trained primates, Travis and Sparks (*1968)* recorded from the globus pallidus of squirrel monkeys trained to perform a go/no-go task. Surprisingly, only 3 of 89 pallidal neurons in their sample displayed activity related to gross motor movements, suggesting that they were not sampling the skeletomotor circuitry within GPi and may instead have been recording from regions with outputs to prefrontal cortex (Travis and Sparks, 1968; Middleton, 2003). Indeed, more than 40% of their task-related cells displayed changes in activity during the instruction period preceding the movement in the go tasks or during the period after the movement prior to the delivery of the reward (Travis and Sparks, 1968; Middleton, 2003).

In summary, the physiological properties of many neurons in the striatum and pallidum of humans and nonhuman primates appears to be similar in many respects to those reported in studies of the physiology of the prefrontal cortex. In addition, compared with the prefrontal cortex, there appears to be an increase in the relative proportion of neurons in the basal ganglia whose activity is significantly influenced by the expectation of reward (Middleton, 2003).

Thus, although the concept that parallel basal ganglia circuits can subserve different types of tasks appears valid, there clearly appears to be another system operating across circuit boundaries that integrates information about the rewarding value of a behavioural act. Shultz *(1998)* proposed that the neural substrate far this reward system is the phasic activity of dopamine synthesizing cells in the SNpc, which conveys information about primary reinforcement and behavioural state to cells in the striatum. Because such influences exist throughout the striatum, it is quite likely that many, if not all, basal ganglia circuits utilize reward-related information to modify the properties of cells within them in order to carry out meaningful behavioural acts (Schultz 1998; Middleton, 2003).

In two above mentioned functional studies, Jueptner and colleagues (1997 A-B) listed significant activations in areas 9 and 46, dorsolateral portions of the caudate, rostrodorsal portions of the globus pallidus, and thalamic sites in VA and paralaminar MD in subjects during the learning of new sequences of eight key presses, compared with activations seen during previously learned sequences. Thus, nearly all of the brain regions that are thought to participate in basal ganglia circuits with the dorsolateral prefrontal cortex were activated during a sequence-learning task in humans (Jeuptner et al., 1997 A and B; Middleton, 2003).

In another PET study, Owen and colleagues (1997; 1998) compared the activity of normal subjects and Parkinson's disease patients during the performance of a difficult planning task (the Tower of London task), a spatial working memory task, and a simple visually guided movement task. Previous studies from this group had strongly implicated areas 9 and 46 in the performance of both the planning and spatial working memory tasks (1996). These investigators found that Parkinson's disease patients had very little GPi activation during the cognitive tasks, while normal subjects had greatly increased GPi activation during the same tasks. The amount of this differential activation was directly correlated with task difficulty, i.e., the greatest differences occurred in the tasks with the most cognitive demands, whereas there were no differences in GPi activation between normal and parkinsonian subjects in the simple motor task (Owen et al., 1996; 1997; 1998; Middleton, 2003). Because patients with Parkinson' s disease had previously been shown to be impaired in the performance of the same cognitive tasks (Owen et al., 1992; Middleton, 2003), the authors concluded that the GPi outputs to the dorsolateral prefrontal cortex (areas 9 and 46) played an important role in these tasks.

Finally, a number of recent studies have examined the functional activation of the basal ganglia and prefrontal cortex during cognitive tasks (mainly card gambling games) that have an artificial reward component (Delgado et al., 2000;Elliott et al., 2000 ). Delgado et a l., (2000) obtained strong evidence for the involvement of the striatum, bilateral pallidum, and ventral anterior thalamus in context-dependent responses to increasing levels of rewards (Delgado et al., 2000; Middleton, 2003).

In: Basal Ganglia and Thalamus
Editors: R. Moretti, P. Torre and R. M. Antonello
ISBN 978-1-60692-198-2


*Chapter 7*

# Behavioural Conditioning Exerted by Basal Ganglia

***Rita Moretti, Paola Torre and Rodolfo M. Antonello***

According to different experimental studies, above described, basal ganglia seem to exert a conditioning properties on behaviour and social conduct.

There are two well-understood in vivo models of pathologies that involved basal-ganglia, directly and via their neural network to limbic-amygdala circuits and prefrontal cortex: these conditions are the post-traumatic stress disorders and the obsessive-compulsive disorders.

Posttraumatic Stress Disorder (PTSD) is one of few psychiatric disorders for which the etiology is defined. The cardinal triad of features:

1) re-experiencing phenomena (e.g., flashbacks),
2) hyperarousal (e.g., exaggerated startle)
3) avoidance or numbing, arise in the aftermath of an emotionally traumatic event (Rauch, 2005).

The disorder can be chronic and debilitating.

Models of PTSD have focused on the interaction among three key elements of the limbic system: the amygdala, medial prefrontal cortex (rostral anterior cingulate cortex - rACC), the head of the caudate, and hippocampus. Initially such models were motivated by extrapolation from animal research on fear conditioning. Fear conditioning studies in rodents have demonstrated that the amygdala plays a central role in the acquisition of conditioned fear as well as mediating the expression of fear responses (see data and literature in Rauch, 2005). Conditioned fear responding is modulated top down by cortical regions in important ways. Extinction of fear responses, and specifically the capacity to retain and recall extinction memory, is mediated by infralimbic cortex in rats, whereas the hippocampus plays a role in providing information about safe vs. dangerous contexts. Thus neurocircuitry models of PTSD have proposed exaggerated amygdale responses (underlying indelible formation of fear associations and excessive responses to reminders of the traumatic event), deficient medial prefrontal cortical function (underlying impaired extinction), and deficient hippocampal function (underlying an impaired ability to appreciate safe contexts) (see data and Literature in Rauch, 2005).

Neuroimaging methods have been used to systematically test the elements of this model. Symptom provocation studies suggest exaggerated amygdala responses to reminders of traumatic events, whereas activity within medial prefrontal/rACC is attenuated in PTSD. Cognitive activation probes have been used to test the functional integrity of the various nodes of this circuit. Experiments using emotionally expressive face stimuli have shown exaggerated amygdala responses to generalized threat-related cues during automatic information processing. In addition, specially designed Stroop tasks have been used with fMRI to show deficient rACC function. Likewise, explicit learning and memory paradigms have been used to show deficient hippocampal function in PTSD. Paralleling these observed functional deficiencies, morphometric MRI studies have further shown selectively reduced volumes in these relevant cortical territories including rACC and subcallosal prefrontal cortex as well as the hippocampus (Bremner, 2003; Rauch, 2005).

In summary, PTSD appears to be characterized by fundamental amygdala hyperresponsivity, that may be exacerbated by insufficient top-down modulation from medial prefrontal cortex / rACC and the hippocampus (Bremner et al., 1995; Gilbertson et al., 2002).

*Obsessive Compulsive Disorder (OCD)* ,categorized among the anxiety disorders, is a condition defined by intrusive cognitions (obsessions) and repetitive behaviors (compulsions) (Rauch, 2005). The concept of an OC-spectrum of disorders, including Tourette syndrome, Body Dysmorphic Disorder and Trichotillomania, has been advanced based on phenomenological similarities as well as high rates of comorbidity. All of these conditions share features of fundamental intrusive events, affective accompaniments, and semi-voluntary stereotyped repetitive behaviors.

The classic heuristic of parallel cortico-striato-thalamo-cortical (CSTC) loops has been leveraged to advance contemporary neurocircuitry models of OCD and related disorders. CSTC circuits normally subserve filtering and amplification functions that are largely automatic or implicit (see data and literature in Rauch, 2005). In the case of OCDs, the implication is that basal ganglia (specifically striatal) pathology leads to impaired inhibition of cortico-thalamic activity, which in turn leads to failures in "filtering", and hence the intrusive events (Graybel and Rauch, 2000).

Though this generic profile may extend across the various OC-spectrum disorders, the specific clinical presentation for a given individual may be governed by the topography of striatum involved (Rauch, 2005). Hence this conceptualization has been dubbed "the striatal topography model of OCDs" (Miguel et al., 1997). In the case of OCD the striatal territory involved is principally the ventral caudate nucleus, linked with orbitofrontal and anterior cingulate cortex (Baxter et al., 1992). Consequently, the intrusive material entails affectively charged cognitions (obsessions).

A broad range of neuroimaging studies have yielded convergent data to support this CSTC model of OCD. Morphometric MRI data and parallel findings of reduced N-acetyl aspartate by magnetic resonance spectroscopy suggest subtle structural abnormalities involving the striatum (Bartha et al., 1988).

Functional imaging studies indicate hyperactivity within the CSTC circuit that is accentuated during symptom provocation and attenuated with treatment.

Of clinical relevance, the magnitude of activity within orbitofrontal cortex predicts subsequent response to pharmacological as well as behavioral treatment. Finally, cognitive activation studies employing an implicit learning task have shown that patients with OCD fail to normally recruit striatum and instead exhibit aberrant

medial temporal activation. This finding parallels the phenomenology of OCD, where affectively charged material that may be put to rest automatically by healthy subjects is instead processed (explicitly) within the conscious domain (as obsessions) (Miguel et al., 1997).

In summary, OCD is characterized by hyperactivity within the elements of a CSTC circuit, namely the orbitofrontal - caudate loop. Though primary pathology may reside within the striatum leading to deficits in filtering and implicit information processing, symptom severity and treatment response may be more closely associated with orbitofrontal cortical activity.

In: Basal Ganglia and Thalamus
Editors: R. Moretti, P. Torre and R. M. Antonello
ISBN 978-1-60692-198-2


*Chapter 8*

# Apathy and Basal Ganglia

***Rita Moretti and Paola Torre***

Apathy is conventionally defined as an 'absence or lack of feeling, emotion, interest or concern'. To clarify the concept of apathy for practical medical purposes, it has been proposed that apathy corresponds to a 'lack of motivation not attributable to diminished level of consciousness, cognitive impairment or emotional distress' (Marin, 1991, 1996). It is unlikely that the concept of 'lack of motivation' represents the underlying mechanism responsible for apathy because it is a projective psychological interpretation of a given behavioral state (Levy and Dubois, 2006).

Indeed, as a syndrome, apathy should be objectively measurable, independently of any psychological interpretation. In this view, apathy is seen as a quantitative reduction of actions compared to the previous behaviour, despite the patient's environmental or physical constraints remaining unchanged. Therefore, one should be cautious when stating that a subject is apathetic if the reduction of actions is contemporary to physical impairments (such as a paresis or an altered consciousness) (Levy and Dubois, 2006). The reduction of actions can be reversed, at least partially, under strong solicitation from the external environment, testifying to a contrast between a deep alteration of self-generated behaviours and a relative preservation of externally driven ones. In consequence, Levy and Dubois (2006) propose to define apathy as the quantitative reduction of self-generated voluntary and purposeful behaviours. It is therefore observable and can be quantified. According to the proposed definition, apathy is a pathology of voluntary action or goal-directed behavior (GDB) and the underlying mechanism(s) responsible for apathy may be seen as dysfunctions occurring at the level of elaboration, execution and control of GDB (Brown and Pluck, 2000).

Apathy is often present after direct lesions of the prefrontal cortex (PFC) (Luria, 1980; Stuss, 2000). It is also a common clinical feature of basal ganglia diseases. It can be observed in neurodegenerative diseases such as Parkinson's disease (PD) (Aarsland et al., 2001; Isella et al., 2002), Huntington's disease (Hamilton et al., 2003; Thompson et al., 2002) and progressive supranuclear palsy (PSP) (Aarsland et al., 2001). Apathy is also frequently encountered after focal lesions of specific structures of the basal ganglia such as the caudate nuclei, the internal pallidum and the medial-dorsal thalamic nuclei (Laplane et al., 1989; Bhatia and Marsden, 1994; Ghika-Schmid and Bogousslavsky, 2000).

Apathy is therefore one of the clinical consequences of the disruption of the PFC-basal ganglia axis, one of the functional systems involved in the generation and control of self-generated purposeful behaviour. From this perspective, a prefrontal-like syndrome (including apathy as one of its clinical manifestations) can be encountered following diseases that mainly involve the basal ganglia. This suggests that apathy can also be the consequence of a 'prefrontal-like' syndrome due to lesions mostly affecting the basal ganglia.

It is thus likely that the physiopathology of apathy is not a single entity but multiple, depending on which specific process or macro-function is disrupted during the completion of GDB. In line with this notion, Stuss et al. (2000) proposed dividing apathetic syndromes into three subtypes: 'emotional', 'cognitive' and 'behavioural'.

Levy and Dubois (2006) observation of patients leads them to replace the term "behaviour" with the concept of 'auto-activation deficit'. This refers to a fundamental deficit of activation of behaviour that is not primarily due to an 'emotional' or a 'cognitive' deficit and can be reversed by external stimulation ('hetero-activation'). This mechanism is associated with the most severe apathetic states. Each of the three groups of mechanisms can be ascribed to lesions of different PFC-basal ganglia territories as follows: 'emotional-affective' to the orbital-medial PFC and presumably to its connected region within the striatum, namely the ventral striatum, 'cognitive' to the lateral PFC (and to its striatal input, namely the dorsal caudate nuclei) and 'autoactivation' to basal ganglia lesions that usually affect both the cognitive and limbic territories [bilateral GPi (internal portion of the globus pallidus) or bilateral paramedian thalamic lesions] and also the dorsal--medial aspect of the PFC (Levy and Dubois, 2006).

In humans, lesions of the orbital-medial PFC are often associated with apathy. For instance, apathy is one of the major signs of frontotemporal dementia (FTD) and is present in more than 90% of patients at the early stage of FTD (mostly affecting the orbital and medial PFC at that stage of the disease) (Pasquier, 1999). After focal orbital and medial PFC lesions, patients often take hours to complete actions that usually take minutes or they remain unable to make a decision or undertake a plan of actions (Eslinger and Damasio, 1985, Levy and Dubois, 2006).

Anatomically, the orbital-medial PFC is connected to limbic (e.g. amygdala, subiculum, ventral tegmental area) and visceromotor (e.g. hypothalamus, periaqueductal gray matter) brain regions (Carmichael and Price, 1995; Ongur and Price, 2000), whereas its most lateral subregion is reciprocally connected to the sensory cortices (primary olfactory, gustatory and somatosensory cortices and association visual and auditory cortices) (Carmichael and Price, 1995; Levy and Dubois, 2006). Altogether, these inputs from limbic structures and from all the sensory systems are very likely the anatomical background that provides the affective or emotional flow of information that is necessary for the orbital-medial PFC to influence the ongoing or forthcoming behaviour (Carmichael and Price, 1995; Ongur and Price, 2000; Levy and Dubois, 2006). Behavioural and electrophysiological studies in the monkey and recent imaging studies in humans indicate that the orbital and medial PFC is essential to provide the contextual value of reward and consequently to integrate the rewarding value of a stimulus into behaviour (O'Doherty et al., 2000, 2001, 2003; Elliott et al., 2003; Levy and

Dubois, 2006). In particular, the influence of the medial-orbital PFC is essential when the context requires the subject to adapt his/her behaviour to maintain a positive outcome and anticipate the rewarding value of a forthcoming behaviour. Electrophysiological studies in the monkey have demonstrated that neurons in the orbital-medial PFC were activated by the contextual (or relative) value of a potential reward rather than by its absolute value: if an animal is hungry, a glucose solution will become attractive and activate orbital neurons. In contrast, when the animal reaches satiety, the self-same neurons no longer discharge (Critchley and Rolls, 1996; Levy and Dubois, 2006).

The orbital and medial PFC projects to the medial and ventral portion of the head, body and tail of the caudate nuclei (Selemon and Goldman-Rakic, 1985; Yeterian and Pandya, 1991) as well as to the 'core' structure of the ventral striatum (Selemon and Goldman-Rakic, 1985). Output from these striatal regions terminates in the medial and ventral pallidum and in the medial portion of the substantia nigra pars reticulata (Haber, 2003).

The pallido-thalamic projections terminate in the magnocellular part of the medial-dorsal (MD) thalamus, which in turn project to the orbital and ventral-medial PFC (Haber, 2003). In addition, this cortico-subcortico-cortical loop is also connected with the amygdala (Fudge et al., 2002) and can thus be considered as a 'limbic loop'. Several physiological studies (cell recordings) in monkeys have confirmed the involvement of the ventral portion of the striatum in a functional axis with the orbital-ventral PFC (Levy and Dubois, 2006): indeed, the ventral striatal neurons also seem to play a major role in integrating the affective or emotional value of a given stimulus into the ongoing behaviour (Hollerman et al., 1998). The main patterns of neural discharge are an anticipatory response to a forthcoming reward and double coding (reward and motor preparation). Taken together, this set of data suggests that:

1) apathy may result from a lesion of the orbital and medial PFC;
2) apathy may in this case be related to a disruption of affective and emotional processing;
3) apathy has not been clearly demonstrated after ventral striatal lesions despite the close anatomical and functional relationship between the orbital and medial PFC and ventral areas of the basal ganglia and the involvement of the ventral striatum in reward processing (Levy and Dubois, 2006).

The 'cognitive inertia', refers to the reduction in goal-directed behaviour due to impairments in the cognitive functions needed to elaborate the plan of actions. It results from impairments in several executive functions that are needed to plan and carry out the purposed action, such as impairments in planning, working memory, rule-finding and set-shifting (Levy and Dubois, 2006).

A cognitive inertia can be secondary to lesions of the lateral PFC due to impairments of executive functions. The lateral PFC, represented by the dorsolateral (BA 9/46), ventrolateral (12, 44, 45, 47) and frontopolar (lateral 10) regions (for reviews, see Goldman-Rakic, 1987; Petrides and Pandya, 1999), is an essential node in the neural network subserving executive functions.

The lateral portion of the prefrontal cortex is tightly connected with the caudate nucleus. There is a gradient from the dorsal to the ventral portion of the caudate nuclei that corresponds to projections from the dorsolateral and ventrolateral portions of the prefrontal cortex respectively (Saint-Cyr et al., 1990; Yeterian and Pandya, 1991;). Selemon and Goldman- Rakic (1985) have shown that these projections extend dorsally along the rostrocaudal axis of the striatum. These anatomical findings support a functionally coherent circuit: lesions of the dorsal portion of the caudate nucleus induce impairments in tasks that are also altered after dorsolateral prefrontal cortex lesions, such as spatial delayed and delayed alternation tasks (Iversen, 1979; Levy and Dubois, 2006). Taken together, these data suggest that the dorsal portion of the caudate nuclei (in particular the head) is a key structure in an anatomical-functional network in combination with the dorsolateral PFC which mostly contributes to executive functions.

Therefore, it is not surprising that, in humans, unilateral or bilateral lesions of the dorsal portion of the head of the caudate nucleus are associated with a massive apathetic syndrome (Richfield et al., 1987; Bhatia and Marsden, 1994; Kumral et al., 1999) in combination with a cognitive inertia, consistently found when behavioural disturbances are present (Kumral et al., 1999).

Taken together, this set of data suggests that:

1) apathy may result from lesions of the lateral PFC and from lesions of the dorsal (associative) territories o of the basal ganglia, in particular lesions of the dorsal portion of the head of the caudate nucleus;
2) apathy may in this case be related to a disruption of cognitive processing that can be called cognitive inertia and which refers to a dysexecutive syndrome, mostly related to difficulties in elaborating new patterns of behaviour (Levy and Dubois, 2006).

Apathy related to an 'Auto-activation' deficit refers to difficulties in activating thoughts or initiating the motor program. Patients with an 'auto-activation' deficit exhibit the most severe form of apathy, characterized by difficulties in self-initiating actions or thoughts ('mental emptiness'), contrasting with relatively spared, externally driven responses (Levy and Dubois, 2006). It seems likely that the 'auto-activation' deficit results from a failure to reach the threshold of initiation/activation of thoughts or actions when subjects should behave on an internal basis but not in automatic response to perception (Levy and Dubois, 2006). This failure of cortical activation is clearly demonstrated by the deep prefrontal hypometabolism observed in PSP and in the auto-activation deficit due to focal basal ganglia lesions (D'Antona et al., 1985; Laplane et al., 1989; Baron, 1994). It is thus possible to propose that the disruption of the prefrontal cortex-basal ganglia-prefrontal cortex loops at the level of the basal ganglia may lead to apathy because basal ganglia processing is no longer able to generate the relevant neural signal at the level of its output targets in the prefrontal cognitive and limbic territories (or in the medial prefrontal cortex) (Levy and Dubois, 2006).

In normal conditions, one may propose that the prefrontal cortex internalizes the information from the external and internal environments needed to make a decision about possible actions to be elaborated and performed. Neural signals corresponding to the thoughts or actions are then processed by the basal ganglia in order to validate the most relevant signal. Validation processing may be translated into the extraction of the relevant signal from noise to be readdressed to the output target, namely the prefrontal cortex (Levy and Dubois, 2006), where a clear-cut signal can be detected and contributes to disambiguating decision-making and maintaining or modifying the ongoing behaviour. In pathological situations, if there is a focal destruction within the basal ganglia sub-regions, the signal emerging from the basal ganglia is diminished, the ongoing behaviour is not validated (i.e. not amplified) at the level of the cortex and could be difficult to maintain, and the forthcoming one (if it is not reflexive) is not activated (Levy and Dubois, 2006). In sum, an 'auto-activation' deficit results from the inability of voluntary thoughts or actions to reach the activation threshold due to a decreased signal-to-noise ratio at the level of the prefrontal cortex (Levy and Dubois, 2006).

If an apathetic syndrome resulting from a focal lesion within the striatum, the globus pallidus or the thalamus can be explained by a disruption of a functional circuit leading to a failure of activation in the prefrontal targets, the same result occurs in Parkinson' Disease (PD), where there are virtually no direct lesions of these structures but rather a cascade of dysfunctions secondary to the loss of striatal dopamine innervation. Classically, dopamine is associated with reward processing (Rolls, 1976; Robbins and Everitt, 1996; Schultz et al., 1997). What emerges from this literature is that dopamine may act as a modulating system that favours reward-dependent learning (Levy and Dubois, 2006).

For instance, dopaminergic neurons may signal discrepancies between the predicted reward as the result of a given behaviour and the reward that the subject eventually receives (Schultz et al., 1997). In addition, dopamine may also code uncertainty of reward delivery (Fiorillo et al., 2003).

Both mechanisms may suggest a possible role for dopamine signals in attention-based learning and evaluating the odds in decision-making based on potentially unpredicted rewarding events. As the reward processing circuit involves the orbital-prefrontal-ventral striatum circuit, the influence of dopamine on reward processing may act through a modulation of this circuit via the meso-cortico-limbic pathway. In consequence, one may hypothesize that apathy in PD patients falls within the subtype of 'emotional affective' mechanisms and may result from a dysfunction of the orbital-medial prefrontal-ventral striatum circuit (Levy and Dubois, 2006).

However, several data suggest that apathy in PD is not due to a disruption of 'emotional-affective' processing:

(I) apathy is present in PD even at relatively early stages and in the absence of dementia (Aarsland et al., 1999; Isella et al., 2002), when the dopaminergic mesocorticolimbic pathway is supposed to be relatively spared (Ruberg and Agid, 1988);

(II) tasks used to assess reward sensitivity and the effect of changes in reward contingencies, such as the gambling (Bechara et al., 1994) and reversal (Rolls et al., 1994) tasks, which are sensitive to lesions of the orbital-medial prefrontal-ventral striatum loop, were not found to be impaired in PD patients, even in patients tested while 'off' levodopa therapy (Czernecki et al., 2002).

According to Levy and Dubois (2006), apathy in PD may result from the disruption of 'cognitive' processing usually mediated by the dorsolateral prefrontal cortex-dorsal caudate nucleus circuit. This is in agreement with the fact that cognitive dysfunction in PD resembles that of patients with direct dorsolateral prefrontal cortex lesions (Pillon et al., 2002) and that dopamine denervation of the alternative target, i.e. the orbital-medial prefrontal cortex-ventral striatum circuit, is significantly less severe.

Another hypothesis is to consider that apathy is one of several clinical syndromes due to the same elementary dysfunction. This proposal follows the suggestion made by Rolls (1989) that : "dopamine is a nonspecific modulator that only sets the thresholds of firing in striatal neurons regardless of what type of information these neurons carry as far as the information is relevant from the cortical perspective". In this framework, apathy can result from the inability of the basal ganglia to validate the relevant signal that is transferred to the prefrontal cortex. In addition, the overactivity of the GABAergic output neurons of the GPi may lead to an excessive and global inhibition of the thalamocortical circuits and thereby hypoactivates the frontal regions involved in the generation of actions based on self-guidance (Rolls, 1989; Levy and Dubois, 2006). In support to this view, several functional imaging studies in non-demented PD patients have shown that while PD patients were asked to perform a freely selected volitional motor task, at 'off ' state, an hypometabolism was observed in the medial frontal cortex (mostly in the rostral supplementary motor area) and in the dorsolateral prefrontal cortex as compared to normal controls (Jahanshahi et al., 1995; Samuel et al., 1997). This hypometabolism was reversed by apomorphine, a dopamine agonist (Jenkins et al., 1992) and by internal pallidotomy (Samuel et al., 1997).

Levy and Dubois (2006) thus hypothesize that, following striatal dopamine depletion, both defaults in selection and signal amplification contribute to apathy because the output structures can no more disambiguate the relevant signal and this may cause problems in decision-making, inducing aborted or delayed responses.

In: Basal Ganglia and Thalamus
Editors: R. Moretti, P. Torre and R. M. Antonello
ISBN 978-1-60692-198-2


Chapter 9

# Parkinson's Disease as a Model of Basal Ganglia Disruption

***Rita Moretti, Paola Torre, Rodolfo M. Antonello, and Antonio Bava****
* Dip. Fisiologia e Patologia, Università di Trieste, Italy

Parkinson's disease (PD) is a common disorder, with close to a million affected patients in the United States. The lifetime risk of developing PD from birth is approximately 1.6% (2% for men; 1.3% for women)(Elbaz et al., 2002). The risk of nursing home placement is almost seven-fold higher among PD patients(PArashos et al., 2002). Although the advent of levodopa therapy markedly reduced mortality rates (Uitti et al., 1993), longevity is still compromised by PD, albeit not dramatically (Elbaz et al., 2003).

There are four cardinal features of PD that can be grouped under the acronym TRAP: Tremor at rest, Rigidity, Akinesia (or bradykinesia) and Postural instability. In addition, flexed posture and freezing (motor blocks) have been included among classic features of parkinsonism, with PD as the most common form. Because of the diverse profiles and lifestyles of those affected by PD, motor and nonmotor impairments should be evaluated in the context of each patient's needs and goals (Jankovic, 2008). Bradykinesia refers to slowness of movement and is the most characteristic clinical feature of PD, although it may also be seen in other disorders, including depression. Bradykinesia is a hallmark of basal ganglia disorders, and it encompasses difficulties with planning, initiating and executing movement and with performing sequential and simultaneous tasks. The initial manifestation is often slowness in performing activities of daily living and slow movement and reaction times. In common with other parkinsonian symptoms, bradykinesia is dependent on the emotional state of the patient. For example, immobile patients who become excited may be able to make quick movements such as catching a ball (or may be able to suddenly run if someone screams ''fire'') (Jankovic, 2008). This phenomenon (kinesia paradoxica) suggests that patients with PD have intact motor programmes but have difficulties accessing them without an external trigger, such as a loud noise, marching music or a visual cue requiring them to step over an obstacle (Jankovic, 2008).

Although the pathophysiology of bradykinesia has not been well delineated, it is the cardinal PD feature that appears to correlate best with degree of dopamine deficiency.

This is supported by the observation of decreased neuronal density in the substantia nigra in elderly patients with parkinsonism regardless of PD diagnosis. In addition, positron emission tomography in patients with PD has demonstrated that the decreased 18F-fluorodopa uptake in the striatum and accumbens– caudate complex is proportional to the degree of bradykinesia. Functional neuroimaging studies also suggest impairment in the recruitment of cortical and subcortical systems that regulate kinematic parameters of movement (eg, velocity) (Turner et al., 2003).

Conversely, recruitment of various premotor areas, such as those responsible for visuomotor control, is increased (Turner et al., 2003).

Anatomically, the deficit appears to be localised in the putamen and globus pallidus (Lozza et al., 2002), resulting in a reduction in the muscle force produced at the initiation of movement.

Rest tremor is the most common and easily recognised symptom of PD. Tremors are unilateral, occur at a frequency between 4 and 6 Hz, and almost always are prominent in the distal part of an extremity. Hand tremors are described as supination–pronation (''pill-rolling'') tremors that spread from one hand to the other. Rest tremor in patients with PD can also involve the lips, chin, jaw and legs but, unlike essential tremor, rarely involves the neck/head or voice (Jankovic, 2008). Clinical–pathological studies have demonstrated that patients with PD and prominent tremor have degeneration of a subgroup of midbrain (A8) neurons, whereas this area is spared in PD patients without tremor.

Rigidity is characterised by increased resistance, usually accompanied by the ''cogwheel'' phenomenon, particularly when associated with an underlying tremor, present throughout the range of passive movement of a limb (flexion, extension or rotation about a joint). It may occur proximally (eg, neck, shoulders, hips) and distally (eg, wrists, ankles). Rigidity may be associated with pain, and painful shoulder is one of the most frequent initial manifestations of PD although it is commonly misdiagnosed as arthritis, bursitis or rotator cuff injury (Jankovic, 2008).

In addition, rigidity of the neck and trunk (axial rigidity) may occur, resulting in abnormal axial postures (eg, anterocollis, scoliosis). Postural deformities resulting in flexed neck and trunk posture and flexed elbows and knees are often associated with rigidity. However, flexed posture generally occurs late in the disease. Striatal limb deformities (eg, striatal hand, striatal toe) may also develop in some patients. Striatal hand is characterised by ulnar deviation of the hands, flexion of the metacarpophalangeal joints and extension of the proximal and flexion of the distal interphalangeal joints; striatal foot is characterised by extension or flexion of the toes (Jankovic, 2008).

Postural instability due to loss of postural reflexes is generally a manifestation of the late stages of PD and usually occurs after the onset of other clinical features. The pull test, in which the patient is quickly pulled backward or forward by the shoulders, is used to assess the degree of retropulsion or propulsion, respectively. Taking more than two steps backwards or the absence of any postural response indicates an abnormal postural response. Postural instability (along with freezing of gait) is the most common cause of falls and contributes significantly to the risk of hip fractures (Jankovic, 2008).

Freezing, also referred to as motor blocks, is a form of akinesia (loss of movement) and is one of the most disabling symptoms of PD (Jankovic, 2008). Freezing most commonly affects the legs during walking, but the arms and eyelids can also be involved. It typically manifests as a sudden and transient (usually,10 seconds) inability to move. This may include hesitation when beginning to walk (start hesitation) or a sudden inability to move the feet during specific situations (eg, turning or walking through a narrow passage, crossing busy streets, approaching a destination). Freezing is associated with substantial social and clinical consequences for patients. In particular, it is a common cause of falls (Jankovic, 2008).

PD is associated with neuronal degeneration in multiple areas of the nervous system. In addition to the prominent loss of nigro-striatal dopamine neurons, there is also degeneration of nor-epinephrine neurons in the locus coeruleus, cholinergic neurons in the nucleus basalis of Meynert, the dorsal motor nucleus of the vagus, and involvement of the spinal cord and peripheral autonomic system. There is dysfunction of multiple neurotransmitter systems and research is underway to determine if therapeutic restoration of these transmitters can provide benefit in PD.

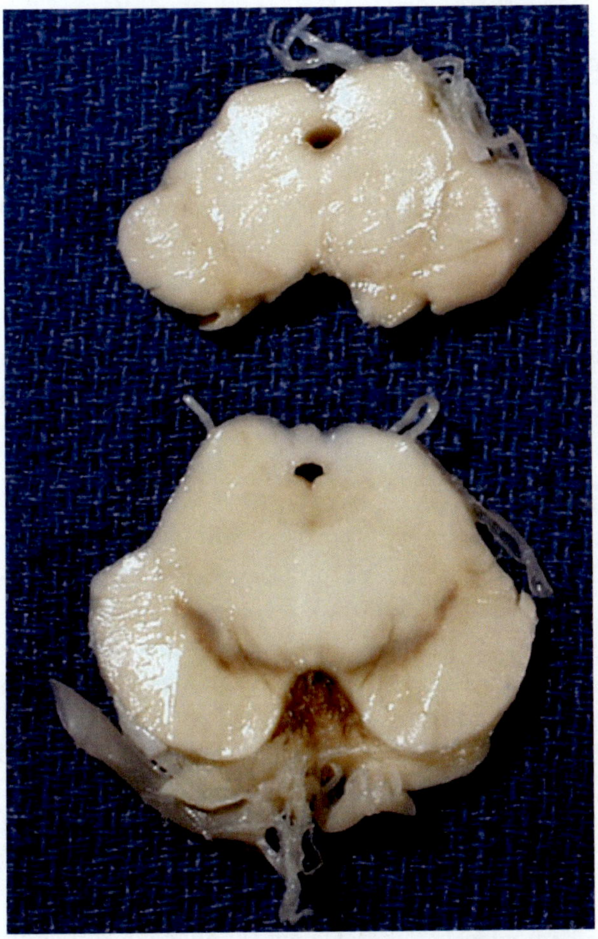

Figure18. Up depigmentation of substantia nigra, in PD patient.Down normal appearance of substantia nigra, coloured parts.

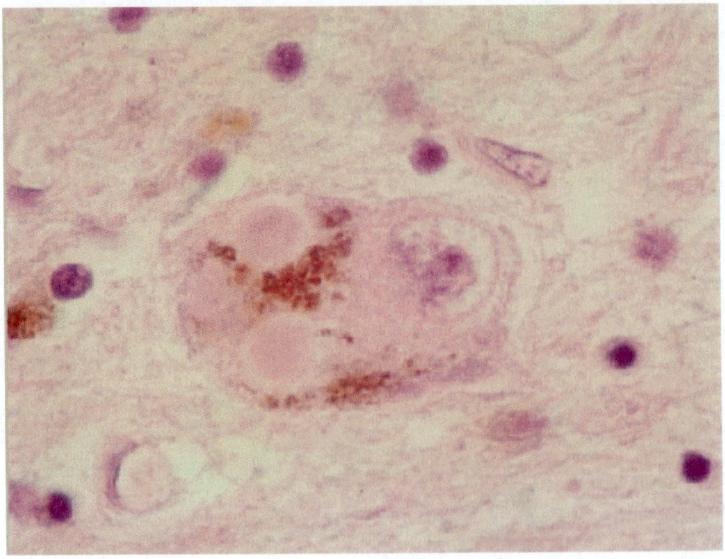

Figure19. Lewy's body in a brain section.

The assumption that the PD neurodegenerative process begins in the dopaminergic substantia nigra has been seriously challenged by recent publications. Braak and colleagues have provided evidence in support of a lower brainstem origin, predating involvement of the nigra(Del Tredici et al., 2002; Braak et al., 2003; Braak et al., 2004). In their six-stage scheme, the substantia nigra does not become involved until intermediate stages (3-4); the final stage (stage 6) manifests as diffuse Lewy body disease, including neocortical involvement. The initial neurodegenerative stage is in the medulla (dorsal motor nucleus of the vagus and adjoining reticular formation) and also the olfactory bulb and portions of the anterior olfactory nucleus(Braak et al., 2003). Whether the autonomic nervous system is also involved at this stage is possible, but uncertain (Braak et al., 2003). This scheme has received support from other neuropathologists (Braak et al., 2004).

Clinical studies have identified several conditions that often precede the motor symptoms of PD by many years.

Although they do not necessarily fit precisely with the Braak stages, they are suggestive. In the prospective Honolulu Heart Study, constipation was associated with a threefold risk of subsequent PD and the mean interval from recorded bowel symptoms to parkinsonism averaged ten years(Abbott et al., 2001). Prospective evaluation of men presenting with isolated REM sleep behavior revealed that 38% went on to develop PD and at a mean interval of 12.7 years (Schenk et al., 1996).

The late stages of the Braak scheme fit with published clinicopathologic studies. Late-developing dementia in PD has the primary neuropathological substrate of proliferation of Lewy body degeneration to limbic cortex and neocortex (Hurtig et al., 2000; Duda et al., 2002). In the most advanced PD, even the motor symptoms have non-dopaminergic substrates(Agid et al., 1990; Duda et al., 2002).

The evolving findings suggest that PD is not simply a dopaminergic neuron disease.

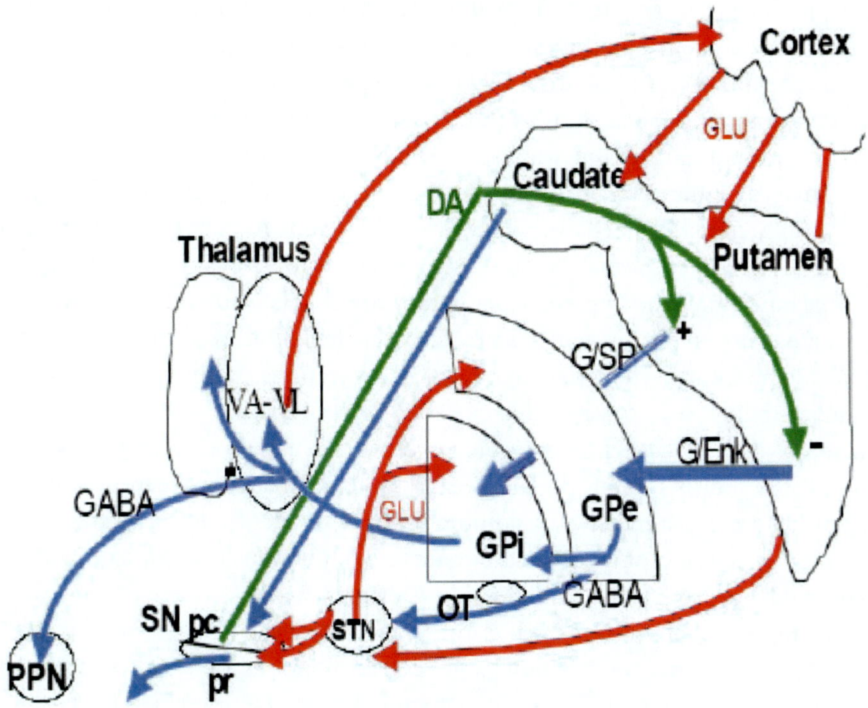


Figure 20. Schematic diagram of the main neuro-anatomical pathways of the basal ganglia.The cortical output is excitatory to striatum and STN; the striatal, GPe and GPi output is inhibitory; the STN and thalamic output is excitatory. The SNpc is dopaminergic.

The earliest stages may not be dopaminergic and the latest stages clearly are not (Hurtig et al., 2000; Duda et al, 2002). This has led to questioning the validity of the previously accepted dopamine oxidative stress hypothesis of PD (Ahlskog, 2002).

Normally, the striatum controls the activity of basal ganglia output neurones via two major routes: the direct pathway and the indirect pathway. Following the loss of striatal dopamine that occurs in Parkinson's disease, these routes fail.It is now known that there are at least five dopamine receptors. The five known receptors can be separated into two distinct families, D1 (made up of D1 and D5) and D2 (made up of D2, D3, and D4).[24] The effect of dopamine and dopamine agonists on D1 receptors is excitatory, whereas their action on D2 receptors is inhibitory. Neurons containing D1 receptors project from the striatum (caudate and putamen) to globus pallidus internus (GPi) and are considered part of the direct pathway. Neurons containing the D2 receptors project from the striatum to globus pallidus externus (GPe) which then projects to GPi via the subthalamic nucleus. D2 receptors are considered part of the indirect pathway.The direct pathway is the simplest pathway through basal ganglia. Striato-pallidal neurones induce an inhibition via gamma-aminobutyric acid (GABA) and substance P upon internal pallidal neurones (GPi) which are themselves inhibitory to target neurones. This leads to disinhibition of the motor thalamus and to an increase in the firing rate of its cells. The proposed effect of activation of the 'direct' pathway is to support or facilitate ongoing movements.

In Parkinson's disease, the depletion of dopamine (acting mainly through D1 receptors - excitatory) from the nigrostriatal projection zones in the striatum (putamen and caudate nucleus), results in the direct pathway being underactive. This leads to decreased firing of thalamic neurones and hence inhibition of initiation of movement.

The indirect pathway is via the subthalamic nucleus (STN). Efferents from the striatum terminate in the external globus pallidus (GPe). Activation of these efferents releases gamma-aminobutyric acid and enkephalin which inhibit the external pallidal neurons. The neurons themselves have an inhibitory projection to the subthalamic neucleus (STN). The STN is therefore disinhibited and its cells increase firing that leads to activation of internal pallidal (GPi) and nigral (SNr) neurones. The output from GPi to thalamus is inhibitory, reducing the excitatory thalamic input to supplementary motor areas. The indirect pathway is thought to inhibit unwanted movement. In Parkinson's disease, depletion of dopamine (acting through D2 receptors - inhibitory) from the nigrostriatal projections in the striatum, results in the indirect pathway being overactive.

Parkinson’s disease is traditionally characterized as a motor disorder resulting from a deficiency of the nigro-striatal dopaminergic system. In recent decades its systematic classification has been extended to incorporate non-motor deficits, including visuo-spatial, visuo-perceptual, visuo-motor, and visuo-cognitive dysfunctions (Levin, 1991).

Current evidence suggests that attentional effects in the primary visual cortex itself underlie some higher level visuo-cognitive functions.

Over the past 20 years our knowledge concerning the rich network of lateral and feedback connections at several levels of the mammalian retina has expanded considerably.

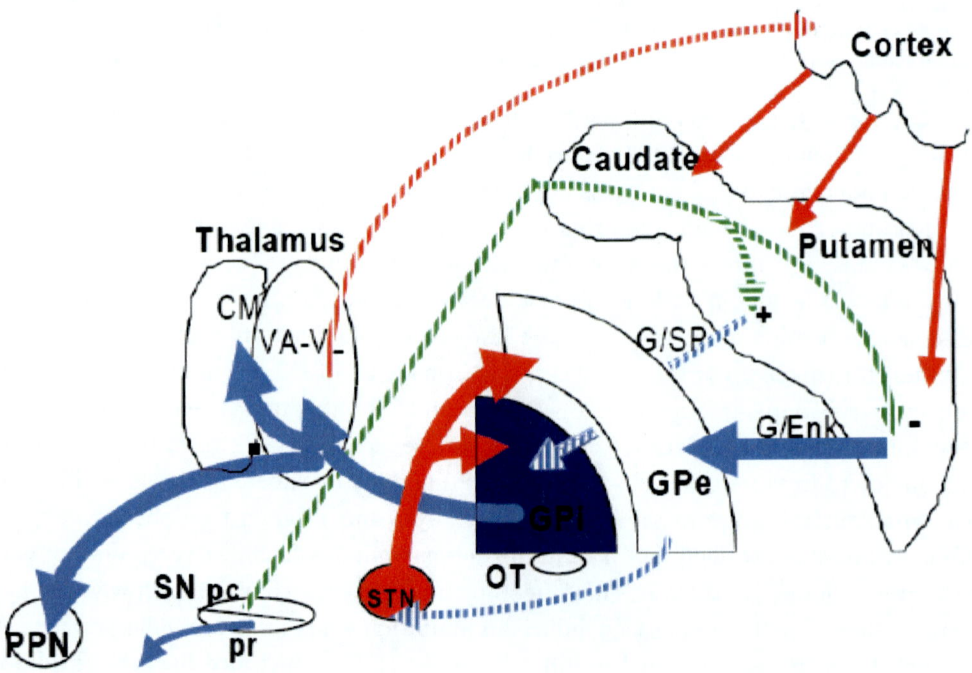


Figure 21. The basal ganglia pathways in Parkinson’s Disease. The thicker lines represent up-regulated pathways. The resulting increased neuronal activity in STN and GPi showed by the diagram occur in PD.

These interactions modify signal processing in the straight-through retinal pathway culminating in the final ganglion cell output. Based on studies in lower vertebrates, it is assumed that the 'surround' in all mammalians is dominated by horizontal cell activity in the outer plexiform layer, while their influence on the 'centre' is negligible (Mangel and Dowling, 1985). Additional lateral interconnections in the inner plexiform layer by amacrine cells most certainly participate in establishing receptive field properties. Various amacrine cell types are distinguishable by neuro-pharmacological means. Dopaminergic amacrine cells are found in all vertebrates, including primates. DA amacrines receive input from other amacrines or directly from bipolars (Hocock and MAriani, 1987; Bodis-Wollner, 2003). In most species DA amacrines act as interplexiform cells, which suggests a role in feedback circuits. Additionally, they may influence in a lateral interconnecting circuit other amacrines in the inner plexiform layer (Bodis-Wollner, 2003). Their contacts are via conventional synapses, however DA amacrines may also connect other neurons through paracrine release of DA. Various DA receptors have been identified in several levels of the mammalian retina, falling into two major classes: D1 and D2 receptors (McGonigle et al., 1988; Denis et al., 1990; Schorderet et al., 1990; Bodis-Wollner, 2003).

The disappearance of the pattern electroretinogram (PERG) has been histologically correlated with retrograde degeneration of the ganglion cells, with hardly any effect on first and second order neurons of the retina preceding ganglion cells. The PERG is spatially "tuned": peak response to sinusoidal gratings occurs with a spatial frequency near the peak of the human psychophysical curve. The current source density profile obtained with full-field and patterned stimulation shows differences in the primate retina (Bodis-WVollner, 2003).

Uniform field stimulation identified both distal and a weaker proximal source, whereas patterned stimuli identified a strong proximal (near the ganglion cell layer) source of the second harmonic response to steady-state stimulation. The results of different experimental sets (Bodis-Wollner, 2003) suggest that the ERG reflects at least two nonsaturating nonlinear response mechanisms in the human retina and one of these precedes the ganglion cells. Dopaminergic amacrine cells form part of the preganglionic retinal organization. Their effect is discernible on preganglionic linear interactions, but their role in non-linear processes has not been explored.

Bodis-Wollner et al. studies (1990), using either 6-OHDA or MPTP showed that the effect of dopamine depletion is a loss of a band-pass retinal spatial response function. However, the change was much more profound using 6-OHDA than MPTP. This difference may be a simple dose-dependent effect or may be the result of subtle differences in the effects of DA neuronal toxins on diverse retinal DA amacrine cells. The effect of the D2 blocker L-sulpiride in low concentration is qualitatively the same as that of DA depletion: peak spatial frequency responses are attenuated (Tagliati et al., 1994).

This similarity suggests that D2 receptors are predominately post-, rather than pre-synaptic in the pre-ganglionic retina. However, we have also observed an attenuation of low spatial frequency responses following high dose sulpiride. With DA depletion we noticed a mild, albeit not significant, increase of the PERG amplitude to low spatial frequency stimuli and the converse and remarkable result was observed with the D1 agonist.

This could occur if low affinity pre-synaptic D2 auto-receptors were involved in controlling neurons predominantly responding to low spatial frequency stimuli. An equal sign effect of the high dose D2 agonist L-sulpiride, and the D1 agonist CY 208-243 argue that low affinity D2 auto-receptors may exist, possibly in the subset of the dopaminergic pathways using D1 receptors, for regulating low spatial frequency responses.

Based on similarity of the results obtained with MPTP and with 6-OHDA, it is likely that they both predominantly impair the type of DA neuron that affects D2 receptors. It is thought that D2 receptors augment coupling between rods and cones in the frog (Wigtovski et al., 1989). Such a mechanism would explain our results, which show that D2 receptor inactivation results in decreasing the gain of the receptive field 'centre' mechanism of retinal ganglion cells.

The damage to this type of DA neuron and/or D2 receptor blockade at low concentrations, causes attenuated peak spatial frequency responses.

A different dopamine dependent pathway modulates horizontal cell coupling via D1 receptors (Piccolino et al., 1984) and dopamine in this pathway uncouples horizontal cells (Teranishi et al., 1983; Lasater et al., 1983).

Damage of neurons affecting this path will allow horizontal cell coupling, dissipating surround signals and weakening the 'surround' organization of retinal ganglion cells. This effect will be especially evident on the responses of neurons with large receptive fields and strong surround organization. Since coupling is under D1 receptor activation, our observations of attenuated low spatial frequency responses under the influence of CY 208-243-247 are consistent with this hypothesis only if low spatial frequency responses are mediated by neurons with strong surround dominance (Bodis-Wollner, 2003).

D2 auto-receptors are involved in the 'surround' D1 dopamine pathway in the primate; blocking them allows a greater DA effect on D1 receptors and enhances 'surround' signals leading to attenuated low spatial frequency responses as we have observed with high dose L-sulpiride. When the surround is weak the neuron becomes a reduced D2 dependent summating device involving retina basal ganglia visual cortex and dorsal visual stream, and frontal-prefrontal circuits (Bodis-Wollner, 2003).

The Baddeley-Hitch model of WM presumes an attentional control system, the central executive, that operates with two slave systems: the phonological loop (PL) (related to the maintaining of auditory and verbal information) and the visuo-spatial sketchpad (VSSP) (related to the maintaining of visual information) (Baddeley, 1998). The VSSP can be further dissociated into a "what" system, providing the basis of object-related visual analysis (e.g., shape, color, and texture), and a "where" system for the processing of spatial location, motion, and three-dimensional properties (Dolan et al., 1997). Several studies provided evidence for a selective impairment of the "where" subsystem of VSSP in early PD: both verbal WM and the "what" visual subsystem seems to be spared until the later stages of illness (Lee et al., 1998; Moreaud et al., 1997; Postle et al., 1997; Bodis-Wollner, 2003). For instance, in visual delayed-response tests, PD patients with mild symptoms were unable to briefly maintain the memory trace of spatial locations of irregular polygons, whereas they successfully kept on-line the shapes of the same stimuli (Postle et al., 1997). PD patients also make significantly more errors in mental rotation of three-dimensional wire-frame figures (Moreaud et al., 1997). In this sense a deviant P300 response to low, but not to medium spatial frequency stimuli may indicate that central processing along pathways for the "what" (medium and higher spatial frequency stimuli) are relatively spared in early PD, while the "where" central pathways receiving information from low and medium spatial frequency stimuli are affected (Bodis-Wollner, 2003).

One possibility of linking primary and cognitive visual deficits in dopaminergic deficiency is that the visual cortex requires a full spectrum of visual input to feed into the dorsal stream of visual processing. A second hypothesis may be that, working memory dysfunction is a consequence of direct dopamine depletion in the prefrontal cortex, reflecting the impairment of the mesocortical pathways (Scatton, 1983). More specifically, transmission mediated by the D1 dopamine receptors may be affected, since these receptors are particularly relevant in the regulation of working memory at both neuronal and behavioral levels (Williams et al., 1995; Muller et al., 1998). Therefore, dopamine deficiency in PD may lead to working memory deficit, and agonists acting on D1 receptors (e.g.,pergolide) could play a useful role in the treatment of parkinsonian cognitive symptoms, but this has not been shown (Bodis-Wollner, 2003).

In: Basal Ganglia and Thalamus
Editors: R. Moretti, P. Torre and R. M. Antonello
ISBN 978-1-60692-198-2


*Chapter 10*

# Deep Brain Stimulation of the Subthalamus: Neuropsychological Effects

***Rita Moretti, Paola Torre, Rodolfo M. Antonello, and Antonio Bava****

* Dip. Fisiologia e Patologia, Università di Trieste, Italy

The basal ganglia were considered as a site for integrating diverse inputs from the entire cerebral cortex and funneling these influences via the ventrolateral thalamus to the motor cortex (Alexander, 1994). The subthalamic nucleus (STN) has generally been considered as a rely station within frontal-subcortical motor control circuitry. The function of the STN in the current model of the basal ganglia organization was developed from animal model and neurodegenerative diseases with hypo- and hyperkinetic movement disorders (Alexander et al, 1986): in this model, the STN represents a relay in the so-called "indirect" pathway of the parallel basal ganglia-thalamocortical circuits. STN excitatory efferents reinforce the inhibition of basal ganglia output nuclei, the internal globus pallidus (GPi) and the reticular substantia nigra (SNr) on thalamo-frontal neurons. A "direct" striato-pallidal pathway opposes the STN by normally inhibiting basal ganglia outflow, facilitating thalamo-frontal drive (Schroeder et al., 2002).

Clinicians maintain firm conviction on the fundamental role of the basal ganglia, and of STN, in motor syndromes such as Parkinson Disease. Lesions at the level of STN lead to hemiballism and involuntary movements of the contralateral extremities (Guridi and Obeso, 2001).

The limits of drug therapy in severe forms of Parkinson's Disease have led to a renewed interest in functional surgery of the basal ganglia and the thalamus. Deep brain stimulation of these structures was developed with the aim of reducing the morbidity of surgery and of offering an adaptive treatment. This type of stimulation was first applied to the thalamus in patients with severe tremor and has recently been applied to the subthalamic nucleus and the internal pallidum. The physiological expectation for this approach is based on the nigrostriatal dopaminergic defect common of Parkinson Disease: i.e., that it induces over-activity of the excitatory glutaminergic subthalamic nucleus-globus pallidus pathway and thus, when the activity is decreased by stimulation, it should lead to an improvement of levodopa-responsive symptoms (akinesia, rigidity, rest tremor, and gait impairment) (Pollak, 2001).

Deep Brain Stimulation (DBS) has become widely used to treat medically refractory tremor, Parkinson's Disease (PD) and primary dystonia. DBS of the thalamus for tremor and of the subthalamic nucleus (STN) and the internal segment of the Globus Pallidus (GPi) for PD is now FDA approved. As the number of DBS procedures for movement disorders, epilepsy and pain increase, more neurologists will need to acquire the knowledge of DBS programming.

DBS comprises chronic high frequency electrical stimulation of the target nucleus, a brain pacemaker, which might work by entraining neuronal firing patterns and pathological synchronization of neuronal activity. The DBS electrode has 4 contacts, which can be used singly in a monopolar mode of stimulation or in combinations of contacts in a bipolar mode. The adjustable parameters of stimulation are the rate or frequency of stimulation, the pulse width, and the voltage. Thus the number of stimulation parameters possible for programming are huge and DBS programming can be a time consuming and daunting task unless a clear protocol is followed. An advantage of DBS compared to ablative procedures is that the stimulation parameters can be adjusted over time, to optimize the clinical benefit while minimizing side effects (Bronte-Stewart, 2003).

As general accordance, the bilateral stimulation of subthalamic nucleus has as a direct consequence a significant bilateral reduction in Parkinson' disability (Kumar et al., 2000). High frequency deep brain stimulation of the STN improves motor function in Parkinson's Disease patients and has therefore become a strategy to treat medically intractable PD patients (Deep Brain Stimulation for Parkinson's Disease Study Group, 2001). Nevertheless, the physiologic mechanism underlying the beneficial effect of deep brain stimulation is not completely understood, but one can argue that it has reversible suppressive effects on the nuclei, and it leads to the hypothesis that high frequency electrical stimulation is neuro-inhibitive, due to the stimulation of GABAergic fibers. In particular, focusing our interest on subthalamic stimulation, there is evidence that, in epileptic animals, modifications in subthalamic nucleus activity can block the ectopical discharges of epileptic foci (Vercueil et al., 1998).

Using functional imaging, it has previously been shown that STN stimulation enhances movement associated activation of supplementary motor area (SMA) (Ceballos-Baumann et al., 1999). Overactivation of the lateral premotor cortex has been shown in an event-related fMRI study investigating joystick movements as well as in block design fMRI studies on complex finger movements (Haslinger et al., 2001; Sabatini et al., 2000). However, facilitation of movement related frontal activity is not necessarily coupled with improvement in frontal executive function as suggested by neuropsychological studies in Parkinson's disease with DBS. Still controversial and debated are the possible consequences on cognition due to DBS-STN, considering that STN role in cognition has not yet been investigated in detail. Clinical studies suggest that the STN participates in non-motor functions which can now be probed in Parkinson's disease patients with deep brain stimulation of the STN, allowing selective and reversible modulation of this nucleus. Bilateral DBS-STN for PD can cause subtle impairments in various aspects of frontal executive functioning, particularly in older patients (Saint Cyr et al., 2000; Jahanshahi et al.; 2000 Pillon, 2002), but results are quite controversial.

In these aspects the striatum may play a more central role, providing important information for subsequent cortical processing.

Though, linguistic aspects have never been strenuously investigated in the subjects with subthalamic nucleus stimulation. There have been attempts to follow some patients with subthalamic nucleus deep brain stimulation (Zanini, et al., 2000; Moretti et al., 2001; Moretti et al., 2002), during which different patients underwent linguistic assessment, before and after surgery. These Authors have found a small, but significant amelioration in linguistic performance during electrical stimulation.

In order to better define the eventual modification of the cognitive profile of patients who underwent to subthalamic nucleus-deep brain stimulation, we have followed nine Parkinson disease patients before and after surgery.

Our study (Moretti et al., 2003- A) included 9 patients (6 men and 3 women) with idiopathic Parkinson' s disease (UK Parkinson's Disease Society Brain Bank Clinical diagnostic criteria) (Hughes et al., 1992). All the patients could be fully studied (mean age 68.7 ± 7.89 years, range= 60-76 years; average age at onset= 48.34 ± 7.89 years, range= 41-56 years). All the patients suffered for 8.91 ± 2.34 years from Parkinson disease and had been treated for several years with levodopa preparations, then developed motor fluctuations and dyskinesias difficult to control by adapting their drug regimen. They were selected as candidates for deep brain stimulation of the subthalamic nucleus because of severe motor fluctuations. In the off medication state, the mean Hohen and Yahr (Hohen and Yahr, 1967) score was 4.01 and the mean Unified Parkinson's Disease Rating Scale (UPDRS II) (Fahn, Elton, and the members of the UPDRS Development Committee, 1987) score was 32.1(activity of daily living) and UPDRS III score was 61.72 (motor examination). The mean preoperative levodopa equivalent dosage was 1432 mg/day. Before surgery, three and eight months later they underwent a neuropsychological evaluation.

All the subjects were right-handed following the criteria of Briggs and Nebes (+ 22.34 ± 1.32, for the Briggs and Nebes test, 1975). Their average educational levels, represented by school years, is of 11.34 ± 5.67 years.

The stereotaxic implant was carried out bilaterally in the subthalamic nucleus, with platinum-iridium electrodes, model Medtronic ® 3389, connected to a subcutaneous impulse generator. Stimuli were provided continuously at 140-180 Hz (square waves of 0.2 ms, voltage less than 3.6 V, intensity 40 μA). Patients responded to neurosurgery, in particular decreasing motor fluctuations and increasing daily living independency.

The stereotaxic implant was carried out bilaterally in the subthalamic nucleus, with platinum-iridium electrodes, Model Medtronic ® 3389. The target visualization was determined using a preoperative stereotactic T1 and T2 weighted MRI. The STN was best identified on the coronal T2 weighted sequences.

The AC-PC line and the midline of the third ventricle were used to define the target coordinates corresponding to the STN in a brain atlas (3 mm posterior to the midpoint of the biscommissural line, 4 mm inferior to AC-PC, 12.5 mm lateral to midline). The electrodes were implanted stereotactically in a single session in accordance (left: x=-13.80; y=8.01; z=

1.75; right: x= 9.87; y=7.71; z= 1.75) with preoperative MRI and intraoperative microrecordings and connected to a subcutaneous impulse generator stimulation. The implanted quadripolar electrodes were positioned as closely as possible to the location where motor benefit was induced by the lowest electrical intensity and adverse effects by the highest electrical intensity using monopolar stimulation. Electrical parameters (pulse width, frequency, and voltage) were progressively adjusted by telemetry, using a console programmer, until an optimal effect was reached, in both the "on" and "off" drug conditions. The stimulation was unipolar, using one contact of the quadripolar electrode. At the time of the study, the mean voltage of stimulation was 2.3 ± 1.02 V; the mean pulse width was of 68 ± 11.34 μs; the mean frequency was 130 ± 23.45 Hz. MRI of the brain was performed the day after electrode implantation to verify the final electrode placement and the lack of significant bleeding. No haemorrhage, infection or ischaemic stroke directly related to the surgery was noted (Moretti et al., 2003- A).

Patients were examined before surgery, after three, six and twelve months after surgery (Moretti et al., 2003-A).

The patients underwent to a Zung self-evaluation for depression (Zung, 1965).

The general cognitive profile was tested by a battery of tests, which comprises: Stroop Test (Stroop, 1935; Perret, 1974), Raven Standard Progressive Matrices (Raven, 1960), digit span backwards and forwards (Wechsler, 1981), retrieval of a story (Wechsler 1945), Syntactic comprehension test and Morphological test (from the Bilingual aphasia test, part B –italian language- Paradis and Canzanella, 1990), alternating verbal semantic fluency (Wechsler, 1981), alternating verbal phonological fluency (Wechsler, 1981), Benton Line Orientation Test (Benton et al, 1983). Moreover, the patients underwent to an alternating syllabic-phonological test, which we have prepared.

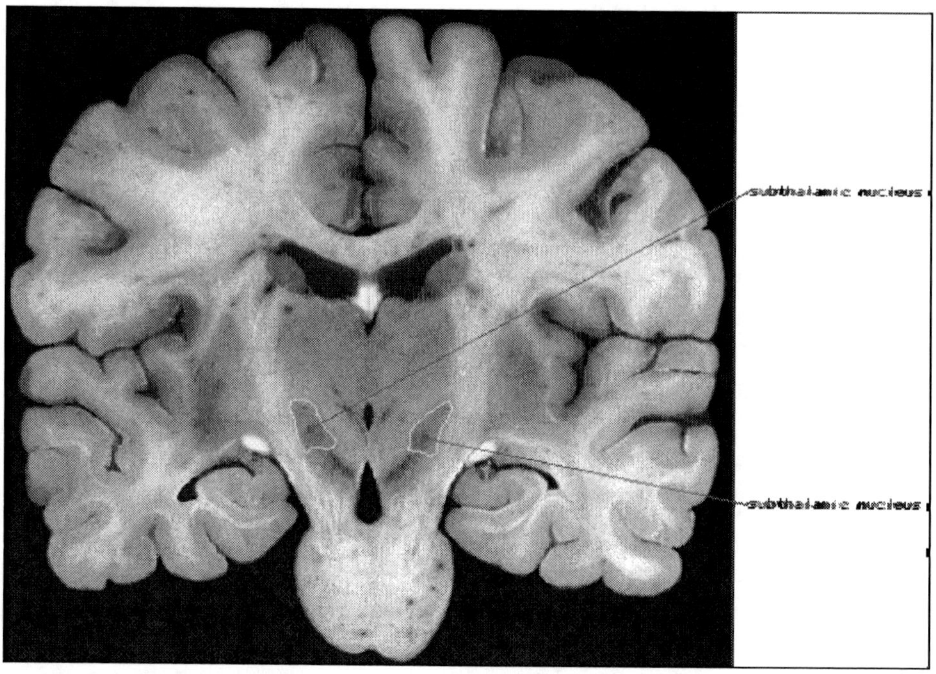


Figure 22. A schematic vision of the subthalamic nucleus.

Subjects were strained to produce in 1 minute time the highest number of words beginning with a given syllable ( "per", "tra", and "di"): we test the total number of words produced and the number of intrusion mistakes (e.g.: if syllabic given target is "tra", we count the number of different nouns beginning with "ter", or "ta", or "tr") eventually produced (Moretti et al., 2002). The order of test' presentation was constant in all the sessions.

The scores obtained by the patients were compared with those obtained by a group of 9 patients (6 men and 3 women) with idiopathic Parkinson' s disease (UK Parkinson's Disease Society Brain Bank Clinical diagnostic criteria) (Hughes et al., 1992) not submitted to DBS, matched for age, educational level, handedness, and disease degree, who previously agreed to participate and gave their informed consent for testing. All the patients could be fully studied (mean age 69.45 ± 7.12 years, range= 60-76 years; average age at onset= 47.64 ± 7.12 years, range= 41-56 years). All the patients suffered for 8.71 ± 1.46 years from Parkinson disease and had been treated for several years with levodopa preparations.

Statistical analyses were performed using the Statistical Package for the Social Sciences (SPSS, version 10.0). Changes from baseline between Groups with DBS and control group were compared using the two-sample Wilcoxon test. This was done for each efficacy variable. Results are presented as means with standard deviations and *p*-values are presented where appropriate (Table 1).

**Table 1. statistical differences obtained on the three trials: a comparison has been made with the results obtained before surgery during the 2nd and 3rd session. Differences from the control population has been made during the 1st session**

| tests | 1st session | 2nd session | 3rd session | controls |
|---|---|---|---|---|
| Stroop | 67.45 seconds ($p<0.05$; $\chi^2$ test); 1 mistakes | 71.24 sec. 1 mistakes (n.s) | 78 sec. 2 mistakes (n.s) | 58.2 seconds; 1 mistake |
| Raven Standard Progressive Matrices (correct) | 36/60 (ns) | 36/60 (n.s) | 36/60 (n.s) | 39/60 |
| Digit span forward (WAIS-R) | 4 ($p<0.05$; $\chi^2$ test) | 6 ($p<0.05$) | 6 ($p<0.05$) | 7.5 |
| Digit span backwards (Wais-r) | 4 (ns) | 3 (n.s) | 3 (n.s) | 4.3 |
| Retrieval of a story (Wais-memory) | 13/22 (ns) | 14/22 (n.s) | 14/22 (n.s) | 17.8/22 |
| Synctactic comprehension (Bilingual aphasia test-b: errors) | 17 ($p<0.05$; $\chi^2$ test) | 14 (n.s) | 14 (n.s) | 11 |
| Morphological tasks (Bilingual aphasia test-b: errors) | 8 (ns) | 8 (n.s) | 8 (n.s) | 8 |
| Alternating verbal semantic fluency (WAIS-R) | 46 | 38 ($p<0.05$) | 37 ($p<0.05$) | 49.32 |
| Alternating verbal phonological fluency | 38 | 26 (ns) | 24 ($p<0.05$) | 42.34 |

**Table 1. (Continued)**

| tests | 1st session | 2nd session | 3rd session | controls |
|---|---|---|---|---|
| Alternating syllabic verbal fluency | 26 (3 interference mistakes) ($p<0.01$ $\chi2$ test) | 21 (3 interference mistakes) ($p<0.05$) | 23 (0 interference mistakes) ($p<0.01$) | 36.9 |
| Benton line orientation test (errors) | 1 (ns) | 1 (n.s) | 1 (n.s) | 0.2 |

Testing before surgery was conducted when the patients were on medication. At baseline, there were no statistical differences between the two groups on any of the scales, a part from an evident increase in time of Stroop execution (72.34 ± 1.42 seconds) in DBS group, when compared to control group ($p<0.05$, paired samples, t-test). No other differences were evident, indicating that the two groups were matched for cognitive performances.

At one-month, patients submitted to DBS showed significant increase in the total time to execute Stroop test, of 5.55 points over baseline ($p<0.01$) and -10.89 over control Group ($p<0.001$), and in number of mistakes in Stroop test, of 3.44 over baseline ($p<0.01$) and –3.34 over control Group ($p<0.01$); there was a significant decrease of correct recall, of –2.3 over baseline (p0.05), of total words produced in semantic fluency tasks, of –7.19 over baseline ($p<0.001$), and of –5.89 over control ($p<0.001$), in phonological fluency tasks, of –3.9 over baseline ($p<0.01$), and of –3.27 over control ($p<0.01$), in syllabic fluency, of –8.3 over baseline ($p<0.001$), and of –9.6 over control ($p<0.001$); on the contrary, there was a significant increase in number of intrusion mistakes, of 5.78 over baseline ($p<0.001$), and of -5.89 over control group ($p<0.001$).

At six-month, patients with DBS showed significant increase in the total time to execute Stroop test, of 12.1 points over baseline ($p<0.001$) and -16.11 over control Group ($p<0.001$), and in number of mistakes in Stroop test, of 2.33 over baseline ($p<0.01$) and –2 over control Group ($p<0.01$); there was a significant decrease of correct recall, of –1.6 over baseline ($p<0.05$), of total words produced in semantic fluency tasks, of –10.97 over baseline ($p<0.001$), and of –9.23 over control ($p<0.001$), in phonological fluency tasks, of –6.2 over baseline ($p<0.01$), and of –6.67 over control ($p<0.01$), in syllabic fluency, of –10.52 over baseline ($p<0.001$), and of –9 over control ($p<0.001$); on the contrary, there was a significant increase in number of intrusion mistakes, of 4.34 over baseline ($p<0.001$), and of –4.56 over control group ($p<0.001$).

At twelve-month, patients with DBS showed significant increase in the total time to execute Stroop test, of 16.33 points over baseline ($p<0.001$) and -20.11 over control Group ($p<0.001$), and in number of mistakes in Stroop test, of 1.88 over baseline ($p<0.01$) and –1.44 over control Group ($p<0.01$); there was a significant decrease of total mistakes in syntactic comprehension tasks, of –3.12 over baseline ($p<0.01$), and of -3.11 over control ($p<0.01$), of total words produced in semantic fluency tasks, of –10.63 over baseline ($p<0.001$), and of –9 over control ($p<0.001$), in phonological fluency tasks, of –5.87 over baseline ($p<0.01$), and of

–6.34 over control ($p<0.01$), in syllabic fluency, of –10.41 over baseline ($p<0.001$), and of –8.89 over control ($p<0.001$); on the contrary, there was a significant increase in number of intrusion mistakes, of 1.34 over baseline ($p<0.01$), and of –1.56 over control group ($p<0.05$).

Subthalamic stimulation has been shown to present an advantage as it induces a higher antiakinetic effect and has positive effects on all parkinsonian symptoms (Dujardin et al., 2001; Iansek et al., 2002).

The subthalamus has generally been considered as a rely station within frontal-subcortical motor control circuitry. It has connections to the substantia nigra, pars reticulata, output pathways from the basal ganglia (Jahanshahi, et al., 2000): the dorsolateral prefrontal cortex is the final target of the thalamic projection from the substantia nigra pars reticulata. A PET study (Limousin et al. 1997) showed that there was a greater activation of dorsolateral prefrontal cortex during effective stimulation of the subthalamus nucleus. Direct cortico-STN projections and indirect connections, via the nucleus accumbens and ventral pallidum from prelimbic and medial orbital cortex, have an excitatory influence on the STN (Maurice et al., 1998).

On the other hand, subthalamic nucleus may be connected with the frontal-caudate-pulvinar circuits, and therefore be involved in verifying the semantic acceptability of the linguistic production and in the so-called language planning-loop (Wallesch and Papagno, 1988; Cummings, 1995; Fabbro et al., 1997). Information processing in the circuits is not strictly sequential, and it seems to control the preparatory linguistic activity (Cummings, 1995).

Having accepted that bilateral stimulation of the subthalamic nucleus has as a direct consequence a significant bilateral reduction in Parkinson's disability, the most interesting aspect of deep brain stimulation of the subthalamic nucleus is the possible consequence on cognitive process, extending basal ganglia's role from motor to cognitive control. Different authors sustained that there was no significant change in memory or executive functions in a sample of 62 patients treated by bilateral subthalamic nucleus or globus pallidus stimulation (Ardouin et al., 1999) when compared to results obtained before surgery, however other authors obtained previously opposite results (Troster et al., 1997) on the same points. Two recent studies have attempted to study cognition in deep brain stimulation. The first (Pillon, 2002) showed that in "on" phases, deep brain stimulation of the subthalamic nucleus patients had no cognitive deficit at 12 months, except for a decrease in lexical fluency (not explicated by the authors), and no differential effect associated with subthalamic nucleus or globus pallidus stimulation. The second study (Jahanshahi et al., 2000) assessed the effects of bilateral deep brain stimulation on subthalamic nucleus and globus pallidus on executive function, assessed with Reitan Trail Making test or on Wisconsin Card Sorting Test (data and literature in: Jahanshahi, et al., 2000; Schubert et al, 2002). The results showed a significant better performances obtained in subthalamic nucleus stimulation, when compared to globus pallidus stimulation, in particular of executive function. The conditional associative learning, and in general change in performance became more erroneous too. Thus, the significant deterioration in learning of arbitrary conditional associations lends some support to the proposal of Marsden and Obeso (Marsden and Obeso, 1994) that following subcortical surgery, patients with Parkinson's disease should exhibit some deficits in situations that require changing behaviour in novel context. In these aspects the striatum may play a more central role, providing important information for subsequent cortical processing.

As demonstrated by Schroeder et al. (2002), DBS in the STN decreased cerebral blood flow responses in the anterior cingulate cortex during the Stroop task, while impairing task performance at the same time. Decreased activation was also located in the ventral striatum, the structure receiving input from the anterior cingulate cortex in the current model of basal ganglia organization. The angular gyrus is associated with encoding of words (Menard et al., 1996; Peterson et al., 1999). The increased activation during stimulation therefore most likely corresponds to difficulties in suppressing the habitual response, i.e. word processing, when patients had actually to name the font colour of the colour word. Alternatively, during off stimulation, anterior cingulate cortex might effectively modulate language processing in the angular gyrus, resulting in decreased activation compared with the on state. This is in good correspondence to a factorial analysis of cognitive processes in the Stroop task, revealing suppression of language processing to facilitate colour naming (Peterson et al., 1999; Schroeder et al, 2002). Schroeder et al. (2002) results are consistent with the emerging identification of the role of the STN as an important basal ganglia input and output structure not only in the motor, but also in the cognitive/limbic domain.

According to a classical model, STN stimulation should result in facilitation of projections to cortical areas, on the contrary, Schroeder et al. (2002) found the reverse effect, namely impaired cognitive performance coupled to reduced activation in the anterior cingulate cortex circuit. The Stroop task requires a high level of conflict monitoring and response inhibition. The differential impact of STN stimulation on cognitive and motor circuitry supports the two opposite predictions of the so-called paradox of stereotactic surgery in Parkinson's disease (Marsden and Obeso, 1994). The motor system devoid of the important basal ganglia input structure STN could continue to function in routine, predictable and automatic movement. "Releasing the brake" on frontal function with STN stimulation improves aspects of motor functions (Limousin et al., 1995). However, in new and unexpected situations, requiring non-automated behaviour, disruption of basal ganglia input at the level of the TN may lead to inflexibility of mental and motor responses, as shown in the Stroop task.

Thus, we decided to study these aspects of cognitive process, in particular relating them with linguistic process. Therefore, our patients underwent to three different alternating test of verbal fluency, which could give us information on our patients capabilities to adapt themselves in linguistic situations that require changing behaviour and to implement effective strategies to select and produce different words with a given root (in our case a syllable). Attention, working memory, and speediness are indispensable to a good performance on this test. In addition to these aspects, some of the other processes that are necessary for performance of this task are: intrinsic generation of responses partly through suppression of habitual responses, monitoring of one's output and switching attention between different stimulus in a set. These are all processes that are strongly dependent on the integrity of the dorsolateral prefrontal cortex.

Our previous work (Moretti, et al., 2001; Moretti et al., 2002) confirmed a general stability of cognitive performances, with a decrease in fluency tasks (confirming the findings of other studies, such those of Troster, et al., 1997 and those of Ardouin, et al., 1999, in the traditional semantic or phonological tasks) associated with a significant reduction of interference errors in syllabic task. This data seem in accordance to what has been found in literature: in addition to being a potential site of storage of learned habits or procedures (Packard and Knowlton, 2002), the basal ganglia also appears to be able to switch between various tasks depending on the demands present. This process is also known as the ability to shift cognitive set in response to the environment. Patients with damage to basal ganglia circuitry exhibit performance costs that are greater than controls when switching from one task to another (Lawrence et al., 2000). In addition, neuroimaging evidence suggests that frontal corticostriatal loops are activated when subjects must select between many possible responses (Desmond et al., 1998). Observations of these stimulated patients supports the prior research; verbal production is decreased in semantic and syllabic tasks, along with an increase in time required to execute the Stroop test, but the quality of production is better preserved, as showed by a significant reduction of interference mistakes in syllabic tasks, and a concomitant reduction of errors in Stroop execution. Mutatis mutandis, syllabic tasks might be considered as a suppression language process task, tightly bound to attention, selection and verbal production. Defects in focal attention has been described as a consequence of the disruption of the fronto-striatal circuit, which can cause a significant impairment in visual search (Chari, et al., 1996). Therefore, better execution on an attention task (Stroop) may be in part dependent on the "releasing the brake", as Jahanshahi said, over frontal cortical function during subthalamus stimulation (Jahanshahi et al., 2000). This situation may be caused by a possible disinhibition of thalamic nuclei, or may simply be related to a facilitation of prefrontal cortical activation. Thus, it appears there is a reduced linguistic output corresponding to a more selective and precise performance.

Chronic stimulation may help restore of the control property on language exerted by deep brain circuits or may implement prefrontal cortex. The result may be consistent with a slowness of cognitive activity, with a reduction of quantitative production, but with an increase of control on the production, which may be more precise and definite. Long term effects on cognition and on language, in particular, need to be studied, and possible different sequel could emerge. This is an issue that requires longitudinal follow up of patients with chronic deep brain stimulation, in order to assess and verify the previous hypothesis.

Few patients were involved in the study, and considering the variability among Parkinson patients, the potential for generalization is seriously limited: though, the results obtained confirm the follow-up previously published (Moretti et al., 2002).

Many different reports seem to evidence the potential side-effect of DBS-STN on cognition, behaviour and social conduct of patients who undergo to this surgery.

It has been accepted that deep brain stimulation (DBS) may help cases of PD where pharmacological therapy could not manage the symptoms any more. Cognitive decline may follow deep brain stimulation (DBS) of the subthalamic nucleus (STN)Changes in cognition between stimulation on and off have also been described(Saint-Cyr et al.,2000; Dujardin et al., 2001; Witt et al., 2004) .

We have described a 68-year-old right-handed man, with a 14 – year history of PD, who had severe rigidity, severe akinesia, freezing of gait, and moderate tremor while resting, despite 1000 mg of L-dopa, and 2.1 mg of pramipexole daily, and disabling dyskinesias, underwent DBS-STN (Moretti et al., 2003-B).

His UPDRS III (Fahn et al., 1987) in off was 49, and in on is 25. Brain MRI was normal. The stereotaxic implant was carried out bilaterally in the STN, with platinum-iridium electrodes (Medtronic ® 3389). The electrodes were implanted stereotactically in a single session in accordance with preoperative MRI and intraoperative microrecordings. The stimulation was unipolar, using one contact of the quadripolar electrode (2.0 V, 90 μs, 130 Hz). MRI of the brain was performed the day after electrode implantation to verify the final electrode placement and the lack of significant bleeding.

After surgery the patient interrupted all the oral therapy: his UPDRS-III in off (DBS –) was 35; in off, (DBS +) was 14. Stimulation status was significantly associated with a better motor score in tremor, rigidity, bradykinesia, gait and postural stability. The general cognitive profile was tested by a battery of tests, as seen in table 1. One month and after five months, we resubmitted the patient to further evaluation. The scores obtained were compared with those of a group of 15 volunteers (data have been presented before). In this period, language was fluid, without semantically deviant sentences. Auditory and reading comprehension was maintained, and repetition skills were preserved. One month after surgery, a CT of the head was normal. The patient reported a difficulty of initiate his speech. Initial stuttering is very limiting for the patient, who, began a target word three or four time, repeating the first phoneme, or the initial syllable, but not being keen of proceeding. The hesitation was not correlated with "freezing of speech". He knew what he had to say (since he could point to, or indicate in a multiple choice, either visually either verbally presented, the defined word and his verbal working memory was well preserved). Automatic speech was conserved (both when the patient cursed to his misfortune, or when he pronounced automatic expression). Spared were his writing and reading capabilities. The situation remains stable five months later (Table 2). A single study reported one case of worsening of dysartria and gait freezing which had existed before the implantation (Thobois et al., 2002). To our knowledge, speech start-hesitation has not been previously described as a consequence of DBS-STN (Moretti et al., 2003-B).

**Table 2. Neurolinguistic evaluation of 5 minutes spontaneous speech. A comparison with healthy control**

| | 1st session | 2nd session | 3rd session | Control |
|---|---|---|---|---|
| Number of utterances | 29 | 28 | 26 | 32 |
| Total word's number | 345 | 231 | 267 | 478 |
| Words/minute | 78 | 56 | 61 | 92 |
| Types | 123 | 109 | 76 | 137 |
| Neologism | 0 | 0 | 0 | 0 |
| Semantic paraphasia | 0 | 0 | 0 | 0 |
| Verbal paraphasia | 0 | 0 | 0 | 0 |
| Substitution of free grammatical mormorphemes | 0 | 0 | 0 | 1 |
| Omission of Free grammatical morphemes | 0 | 0 | 0 | 0 |

| | 1st session | 2nd session | 3rd session | Control |
|---|---|---|---|---|
| Circumlocutions | 2 | 4 | 5 | 1 |
| Perseverations of words | 1 | 9 | 11 | 0 |
| Semantically deviant sentences | 0 | 5 | 2 | 0 |

Trying to make some speculation, start-hesitation ( by definition that of gait) is a typical initial symptom of the freezing phenomenon (Fahn and Quinn, 2000). Our patient, who had freezing of gait before surgery, manifested speech start hesitation, without other motor symptoms, after DBS-STN. One might say that the mechanism of action of DBS remains to be defined (Ashby, 2000). By whatever mechanism, stimulation mirrors the effects of a destructive lesion.

Subthalamus may be related to speech control by its connections to the substantia nigra, pars reticulata, output pathways from the basal ganglia (Ojemann, 1982): the dorsolateral prefrontal cortex is the final target of the thalamic projection from the substantia nigra pars reticulata. A PET study (Ojemann, 1982) showed that there was a greater activation of dorsolateral prefrontal cortex during effective stimulation of the subthalamus nucleus. With the cortico-striato-thalamo-cortical loop the cortical structures of all cerebral lobes can control the cognitive – motor activity of the frontal lobe structures. The latter are involved in the generation and coordination of voluntary movements (Ojemann, 1982), and maybe in language control. Echolalia, i.e., additional characteristic disorders presented in left basal ganglia lesions, results from the inability of the speaker to inhibit the motor internalisation of verbal expressions produced by the patient (Ojemann, 1982). Chronic stimulation may help to restore the control property on language exerted by deep brain circuits or may implement prefrontal cortex. The result may be consistent with a slowness of cognitive activity, with a reduction of quantitative production . May we say that speech start-hesitation is an excess of control? If so, why has this control been exercised in speech output and not in motor act? (Moretti et al. 2003)

Smeding et al (2007) describe a patient with severe decline after STN DBS and postoperative electrode displacement. The patient is a 43-year-old woman diagnosed with Parkinson disease 8 years prior to DBS STN for severe response fluctuations and dyskinesias. Neuropsychological evaluation was normal. She used levodopa/carbidopa slow release 400/100 mg daily. Surgery was performed using MRI, four-tract microrecordings, and macrostimulation for target localization (12 mm lateral, 2 mm posterior, and 4 mm inferior to the midcommissural point [MCP]). The left electrode (model 3389; Medtronic) was implanted in the central trajectory, with the deepest contact 7 mm below MCP. There was considerable CSF leakage from the burr holes. On the right side, no typical STN activity could be recorded. Because test stimulation along the medial trajectory induced a large reduction in rigidity and bradykinesia, the DBS electrode was implanted with the deepest contact 4 mm below MCP (Smeding et al., 2007). Postoperatively, a stereotactic CT scan showed intracranial air bifrontally with posterior and caudal brain shift. The electrode contacts were at the intended position. Confusion resolved within 1 day. At discharge, the stimulation settings were monopolar, contact 0, pulse width 60 microseconds, frequency 130 Hz, and on the right side amplitude 1.8 V and left side 1.3 V. The levodopa/carbidopa slow-release dosage was reduced to 200/50 mg; 2 mg of pergolide was added. Six months after surgery, motor functioning was satisfactory. Neuropsychological testing showed decline of verbal memory, selective and divided attention, and verbal fluency Her relatives confirmed increased forgetfulness, word-finding difficulties, slowed comprehension, and increased irritability. One year postoperatively, she had progressive memory decline. Neurologic examination showed no changes in motor functioning. Compared with 6 months after surgery, there was a progressive decline in verbal fluency but not in memory. Subsequently, both neurostimulators were turned off. Neuropsychological examination 1 week later showed complete recovery of her cognitive functioning (Smeding et al., 2007). The neurostimulator settings were changed to monopolar at contact 2. One month later, she had no cognitive complaints and motor functioning was satisfactory. Formal testing demonstrated a decline in verbal fluency, but other scores were comparable with the preoperative state. A new stereotactic CT scan was made and co-registered with the preoperative MRI. On the left side, contact 0 was located 12.5 mm lateral, 2 mm anterior, and 3 mm inferior to MCP, in the internal capsule lateral to the anterodorsal STN. Contact 2 was located at the border of the internal capsule and the zona incerta adjacent to the globus pallidus internus (13 mm lateral, 4 mm anterior, 0.5 mm superior relative to MCP). On the right side, contact 0 was in the dorsomedial globus pallidus externus (17 mm lateral, 8.5 mm anterior, 3.5 mm superior to MCP). Contact 2 was in the dorsolateral globus pallidus externus close to the putamen (17 mm lateral, 10.5 mm anterior, 7.5 mm superior to MCP). Cognitive decline resolved after switching off DBS. Adjustment of stimulation settings to the dorsal contacts led only to a minor setback in cognition (Smeding et al., 2007). The electrodes were probably displaced as a result of brain shift caused by CSF leakage during surgery and subsequent unfolding of the brain in the postoperative period. Since this incident, we close the burr holes during surgery with Tissuecol to reduce CSF leakage. Cognitive decline was worst with the ventral contacts activated, which were located outside the STN, on the left side in the internal capsule close to the anterodorsal lateral STN, and on the right side in the dorsomedial globus pallidus externus. Adjustment of stimulation settings did not lead to an improvement in off-phase parkinsonism measured with the Unified Parkinson's Disease Rating Scale, though there was an evident decrease of dyskinesias and less on–off fluctuations possibly related to stimulation of the globus pallidus (Smeding et al., 2007).

Concerns have been raised about the possibility that subthalamic deep brain stimulation (STN-DBS) could also produce cognitive changes and mood and behavioural alterations. Valldeoriola et al. (2006)report the clinical and neuropathological features of a patient with advanced PD who developed behavioural changes and dementia while on STN-DBS. A 74 year old man suffering from PD for 11 years presented troublesome dyskinesias and unpredictable motor fluctuations that did not respond to multiple changes in medication. The Hoehn and Yahr stage was IV while "off" and III while "on" medication. The Schwab and England scale score was 40% in the "off" and 80% in the "on" condition. The UPDRS-III score was 56 while "off" and 21 while "on". No clinically evident signs of cognitive impairment were present. The Mini-Mental State Examination (MMSE) score was 28/30. Neuropsychological assessment was considered to be normal except for the presence of mild cognitive processing slowness and free recall impairment.

Dyskinesias and motor fluctuations had disappeared 3 months after STN-DBS. The Hoehn and Yahr stage was III and the Schwab and England score was 70%. Dopaminergic medication was initially reduced to 20% but the patient later developed restless legs syndrome (attributed to the reduction in dopaminergic medication) and levodopa was reintroduced (400 mg/day). Motor performance remained stable during the following months. There were no significant changes in neuropsychological performance 6 months after STN-DBS. However, shortly afterwards the patient developed mood changes consisting of apathy, anhedonia without sadness, and diurnal hypersomnolence. Levodopa was then increased and antidepressants were started. Changing stimulation electrical parameters and stimulation poles did not change the patient's mood. The patient developed fluctuating confusion, visual hallucinations, and paranoid ideations 1.5 years after surgery. The MMSE score was 22/30. He became violent with his relatives. Clozapine (100 mg/day) reduced aggression but confusion persisted. The possible role of STN-DBS on these cognitive and behavioural changes was assessed by switching off the stimulators for 1 week. Mental status remained unchanged but parkinsonism worsened until the stimulators were again switched on. Later in the course of the disease, the patient received galantamine which improved psychiatric symptoms and temporal-spatial orientation. The patient died from bronchopneumonia 3.5 years after surgery (Valldeoriola et al., 2006). Macroscopic examination of the brain disclosed important loss of pigmentation in the substantia nigra pars compacta and locus coeruleus. The electrode tips were placed within the boundaries of the subthalamus on both sides. On microscopic examination moderate inflammatory infiltrates of T lymphocytes and mild astrocytic gliosis were observed surrounding the leads. Lewy bodies (LB) and Lewy neurites were found in the substantia nigra pars compacta, locus coeruleus, raphe nuclei, the dorsal nucleus of the vagus, and the hypoglossal nerve. Dystrophic neurites and cytoplasmatic inclusions were found in the nucleus of Meynert and the subthalamus (Valldeoriola et al., 2006). Cortical type LB were observed in the gyrus cinguli, transentorhinal cortex, the amygdala, and the parietal and temporal cortex. The density of LB was maximal in the sub-cortical nuclei and the limbic areas. A few neurons showing neurofibrilar degeneration were found in the transentorhinal and entorhinal cortex, amygdalar complex, locus coeruleus, and raphe nuclei. Neocortical senile plaques were scarce and signs of amyloid angiopathy were absent. The pathological diagnosis was diffuse Lewy body disease, transitional type (Valldeoriola et al., 2006). This report (Valldeoriola et al., 2006) is the second clinicopathological description of a patient with PD, dementia, and STN-DBS. The first report by Jarraya et al (2003) described a parkinsonian patient who had some cognitive impairment before STN-DBS and whose mental condition worsened after surgery. Post-mortem examination confirmed the diagnosis of PD. The authors concluded that the bad outcome after surgery was due to the inadequate selection of a candidate who had cognitive deterioration before surgery, but the cause of the worsening mental status after STN-DBS remained unclear. In the case reported here, post-mortem examination disclosed changes typical of diffuse Lewy body disease. In addition, the cognitive and behavioural alterations were not thought to be related to STN-DBS since mental problems appeared late after STN-DBS and changes in stimulation parameters or even disconnection did not result in any mental or behavioural change. The clinical manifestations, the course of the disease, the presence of hallucinations, and the good response to cholinesterase inhibitors were also typical of so called Parkinson's disease-dementia in which the main pathological substrate is the presence of LB in limbic and neo-cortical areas (Cummings, 2005). The case report (Valldeoriola et al., 2006) supports the view that mental problems observed after STN-DBS in some patients are not related to STN-DBS and suggests alternative explanations. A better understanding of the predictive factors for the development of dementia in PD will help to improve candidate selection for STN-DBS (Valldeoriola et al., 2006).

In: Basal Ganglia and Thalamus
Editors: R. Moretti, P. Torre and R. M. Antonello
ISBN 978-1-60692-198-2


*Chapter 11*

# Liver Diseases and Basal Ganglia

***Monica Asaro****
Dipartimento Medicina Clinica e Neurologia, Università di Trieste,
Medicina Clinica, Italy

Patients affected by cirrhosis often present extrapyramidal features due to deep worsening of basal ganglia circuits, with an increase of dopamine metabolism (see Data and Literature in Burkhard et al., 2003), and consequently a decrease of D2 dopamin receptor density (see Data and Literature in Burkhard et al., 2003), an altered glutamate or γ-aminobutyric acid neurotransmission and a reduced glucose consumption in basal ganglia (see Data and Literature in Burkhard et al., 2003).

Usually the onset of cirrhosis-related parkinsonism is very insidious.

The first symptoms are usually observed by relatives rather than reported by patients. The intervening time from the onset of symptoms to the development of severe parkinsonism is short, ranging from 2 to 18 months (mean 7.2 months). Initial symptoms are global slowness of movement, gate, impairment characterized by short and shuffling steps. Other symptoms at onset might be micrographia, dysarthria, hypomimia. At this stage resting tremor and postural instability are usually not present yet.

Usually severe akinesia and lead-pipe rigidity - contrary to what is observed in patients affected by Parkinson's disease - is symmetric. Oculomotor, cerebellar, pyramidal, or sensory abnormalities are lacking. Cognitive functions are globally preserved except for some degree of frontal lobe disfunction and mild depression. Parkinsonism develops independently of and can be clearly separated from encephalopathy episodes, although porto-systemic shunting and history HE seem to constitute necessary background for parkinsonism to occur (Burkhard et al., 2003).

More than 75% - in some studies even 90% - of cirrhotic patients show a symmetric hyperintensity of the pallidum of both sides of T1-weighted images regardless of the presence or otherwise of neurological symptoms. It has been found that there is a significant correlation between blood manganese (Mn) levels and hyperintensity of globus pallidus, showing that MRI alterations are probably due to the deposition of Mn (see data and literature in Weissenborn et al., 2004).

Normally ingested Mn is rapidly cleared by the liver and excreted in to the bowel. In cirrhotic patients the eliminations of Mn via the bowel decreases. As a consequence of it blood Mn levels rise and an increase of cerebral Mn deposition is observed (see Data and Literature in Burkhard et al., 2003). Mn accumulations is more pronounced in the pallidum, putamen and caudate nucleus.

There is a significant correlation between the so called Pallidal Index – calculated as the ratio of the globus pallidus signal intensity to frontal subcortical white-matter signal intensity on T1-weigthed MRI – and blood Mn levels (see data and literature in Weissenborn et al., 2004).

Researchers have investigated other causes of hyperintensity in T1-weighted imagines. Brain copper concentration in bioptic samples in liver failure patients were measured. Copper, as well as Mn, is normally excreted via the epatobiliary system. Whereas Mn is a paramagnetic metal that shortens the T1-relaxation time sufficiently to produce the pallidal T1 hyperintensity of liver failure, copper typically results in increased T2 signal abnormalities.

Klos measured the concentration of iron and zinc in all brain regions – in bioptic samples in liver failure patients – without noticing evident differences (Klos et al., 2006).

Neuroradiological studies show global alterations in brain function in patients with hepatic cirrhosis.

They demonstrate a redistribution of cerebral flow and metabolism from cortical to brain regions, including the basal ganglia (see Data and Literature in Spahr et al., 2002). In addiction to this cerebral biochemical alterations are observed both in grey and white matter tissue (see Data and Literature in Spahr et al., 2002). The most severe alterations have been reported in basal ganglia (see Data and Literature in Spahr et al., 2002).

Observing RM images, hyperintensity in T1-weigthed once is not only observed in the pallidal but these changes extend to other brain regions such as the limbic system (see Data and Literature in Spahr et al., 2002) and the occipital white matter tissue (see Data and Literature in Spahr et al., 2002). This hyperintensity increases with the degree of liver failure (see Data and Literature in Spahr et al., 2002), the size of portal systemic collateral vessels, and following the creation of portosystemic shunt (see Data and Literature in Burkhard et al., 2003).

According to different studies, hyperintensity of the globus pallidus on T1-weighted MRI is correlated with the creation of portosystemic shunt regardless of hepatic function measured by Child Turcotte Pugh method. Hyperintensity of the globus pallidus in patients with idiopathic portal hypertension (IPH) was higher than that observed in patients with hepatic cirrhosis.

As liver functions are usually preserved in patients with IPH, the results strongly suggest that pallidal hyperintensity on MRI does not reflect liver cell dysfunction but the creation of portosystemic shunt does (see Data and Literature in Burkhard et al., 2003).

To corroborate those findings, Krieger et al. described an increase in pallidal signal intensity in transgiugular intrahepatic portosystemic shunt (TIPS) patients, with subsequent augmentation of the portosystemic shunt volume that increases the systemic Mn load.

At the 6-month follow-up, neurological findings indicated that decline of mental status and motor performance appeared more frequently in TIPS patients than in patients who received elective sclerotherapy. Moreover, high pallidal T1 signals have also been observed in patients with long standing portal-systemic shunt alone, due to portal vein thrombosis, without liver cirrhosis (Bang et al., 2007).

Literature on this latest issue is sometimes in disagreement. For example, according to Bang et al. (2007) there is a significant correlation between blood Mn and pallidal index and between Child Pugh scores and pallidal index. They have otherwise demonstrated that there is no correlation between blood Mn and bilirubin concentrations. Patients with biliary obstruction, but no presence of porto-systemic shunt, had a mean blood Mn and T1 pallidal signals similar to that of healthy patients. Further confirmation of what above stated comes from high pallidal signal reported in children with congenital abnormalities of portal-systemic shunt due to dysgenesis of portal venous system but who showed almost normal liver functions (Bang et al., 2007).

Therefore it follows that the liver plays an important role in regulating the redistribution and excretion of Mn, since portal-systemic shunt probably decreases the hepatobiliary clearance of Mn, resulted in increased blood Mn levels along with high pallidal T1 signals.

To supporte it, it was observed than 1 year after transplantation relaxation time values in basal ganglia and thalamus are significantly higher than values obtained before transplantation and after 1 year there is no statistically significant difference (Herynek et al., 2001).

As hinted, radiological studies have highlighted that other than deposition of Mn, other biochemical alterations might be responsible for typical clinical manifestations in these patients.

Recent studies have shown that Mn exposure decreases glutamate uptake and increases the expression of the glycolytic enzyme glyceraldehydes-3-phosphate dehydrogenase suggesting that it may influence glutamatergic neurotransmission and cerebral energy metabolism in hepato-encefalopathy (Hazell et al. 1999). Like ammonia, Mn also induces an increase in peripheral type benzodiazepine receptor (PTBR) binding sites (see data and literature in Herynek et al., 2001). The activation of PTBR elicits the astrocytic synthesis of neurosteroids which are potent positive modulation at the GABA A receptor complex (see data and literature in Herynek et al., 2001).

Considering the histopatological findings common in all these patients, apart from the deposition of Mn, the most probable mechanism is the presence of regional differences in cerebral blood supply and regional differences in the permeability of the blood-brain-barrier for example for the ammonia.

Contrary to what is usually observed, patients affected by hepatic cirrhosis show the highest CBF in the caudate and lenticular nucleus, the posterior cingulate and the cerebellum (see data and literature in Weissenborn et al., 2004). There is a significant correlation between the regional blood flow in the basal ganglia and the mesial temporal lobe and measures of motor function and speed and complex attentional tasks with an increase of CBF with decrease performance. The permeability of the blood brain barrier for ammonia is nearly doubled in patients affected by cirrhosis compared to healthy patients (see data and literature in Weissenborn et al., 2004).

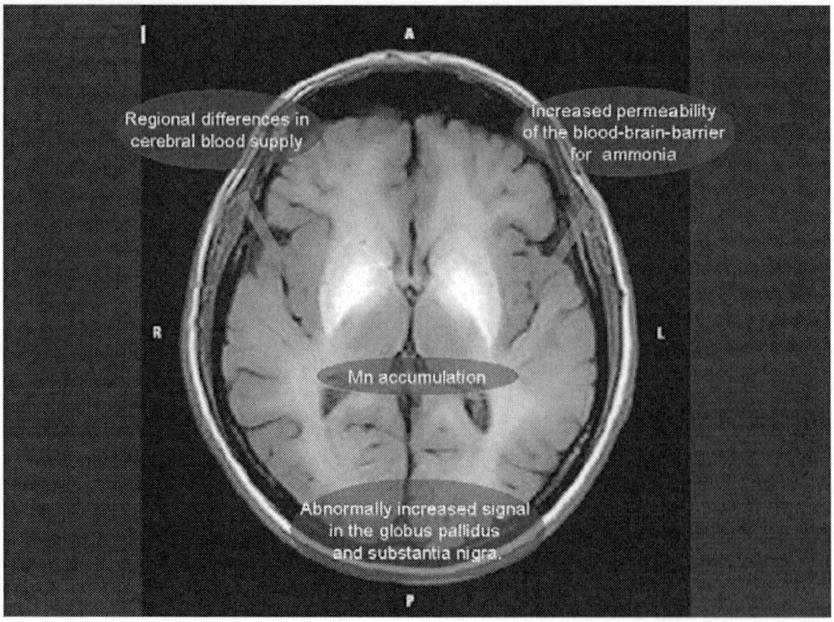


Figure23. MRI alterations of brain structure in chronic liver disease (modified by the Author).

In these latter findings there is evidence that indeed the basal ganglia and also the cerebellum are predominantly exposed to increased ammonia supply (see data and literature in Weissenborn et al., 2004).

MR spectroscopy (MRS) images have allowed us to determine the effect of an increase of ammonia at cerebral level. Ammonia is known to inhibit cellular chloride channel, which contributes to depression of the central nervous system.

Ammonia also facilitates the uptake of tryptophane into the brain, which is a substrate for many metabolites, including serotonin. Therefore, ammonia also decreases glutamatergic neurotransmission causing neurodepression (see data and literature in O'Carroll, 2008).

Patients with cirrhosis and hepatic encephalopathy (HE) show an increase of the glutamate/glutamine signal combined with a decrease of the myo-inositol signal. In addiction the choline signal intensity also seems to decrease in most of the regions but especially in the basal ganglia (see data and literature in O'Carroll, 2008). The MRS alterations are considered to reflect basic metabolic alterations of the brain in cirrhotics that are distinctly involved in the development of HE (see data and literature in O'Carroll, 2008).

The astrocyties accumulation of glutamine is thought to be counteracted initially by the extrusion of myo-inositol. Different Authors have suggested that low grade astrocytic swelling may lead to selective alterations of blood brain barrier function, increased GABAergic tone due to swelling induced augmentation of the synthesis of neurosteroids, alteration of neurotrasmitted processing and glycogen deposition in the astrocytes (see data and literature in O'Carroll, 2008). The extent of the MRS alterations increases with the increase of grade encephalopathy. The main changes however have also been observed in cirrhotic patients with neither clinical nor psychometric or neurophysiological signs of cerebral dysfunction (see data and literature in O'Carroll, 2008). Therefore the MRS alterations cannot be used to make the diagnosis of HE.

As the Glx/Cr and the Myo/Cr alterations are considered to be closely related to the cerebral ammonia metabolism, they could be expected to change rapidly with changes in ammonia supply.

Furthermore it could be expected the MRS changes are more pronounced in those regions that are extraordinarily supplied with ammonia such as basal ganglia. The Glx signal intensity normalizes within 1-2 month after liver transplantation, the myo-inositol intensity within 3-7 month (see data and literature in Weissenborn et al., 2004). Different Authors (see data and literature in Weissenborn et al, 2004) reported that the increase in the Glx concentration in basal ganglia exceeded the change in other regions. These finding are in accordance with the greatest difference in the Chol/Cr and Myo/Cr ratios in the white matter but the largest increase in the Glx/Cr ratio also in the basal ganglia (see data and literature in Weissenborn et al., 2004). It has been observed that there was a decrease in the glucose utilisation of the cortical regions whereas the metabolic rate for glucose of the caudate nucleus, the cerebellum and the thalamus increase (see data and literature in Weissenborn et al., 2004).

Brain abnormalities in patients with HE show that glucose utilization is decreased in cingolum and frontal and parieto-occipital cortex and increased in basal ganglia, cerebellum and temporomesial structures.

As already stated, in patient with HE there is a redistribution of cerebral blood flow (CBF) from the cortex to the thalamus and to the cerebellum. CBF is significantly higher in the basal ganglia and cerebellum than in the cortical areas. Cerebral ammonia metabolism is confined to the astrocytes, where ammonia is metabolized to glutamine. The ammonia-related cellular alterations are considered to be important not only for the development or lethal hepatic coma but also in minimal HE (see data and literature in Ahl et al., 2004). The glial changes correlate with blood ammonia levels and clinical state. Histopathological studies of the brain of patients who died due to liver cirrhosis showed astrocytic alterations mainly in the lenticular and dentate nuclei the cerebral cortex, and the diencephalic and other brainstem nuclei (see data and literature in Ahl et al., 2004).

These histopatological findings correspond to PET finding which showed an increased ammonia supply especially in the basal ganglia and the cerebellum. The increase ammonia supply induces alterations of astrocyte morphology and function, such as cell swelling and altered expression of glutamate transporter and mitochondria peripheral-type benzodiazepine receptor. These alterations are considered the main cause in distinct changes in glutamatergic and GABAergic neurotransmission (see data and literature in Ahl et al., 2004).

Concentration of 3-methoxytyamine (3-MT) increased in the caudate nucleus, whereas concentration homovanillic acid (HVA) increased in the frontal cortex in the caudate nucleus and the putamen. Noradrenalin levels (NA) were greatly increased in the pre-frontal and frontal cortex but their levels remained unchanged in the caudate nucleus. No significant differences in the concentration of either serotonin and its precursor 5-hydroxytyptophan were found.

However 5-hydroxyindoleacetic acid, the major metabolites of 5-HT was increased in the prefrontal cortex and in the caudate nucleus of cirrhotic patients.

These findings show that selective alterations of cathecolamine and 5-HT systems are involved in human HE and therefore they may play an important role not so much in the portosystemic encephalopathy but more in extrapiramidal symptoms (Bergeron et al., 1995; Bergeron et al., 1989).

From a clinical perspective some patients may suffer from abnormal movement such as quick twitches of the facial muscles or the fingers or arms or grimacing. Moreover variable degrees of involuntary movements, which sometimes resemble athetosis, can also be observed, most prominently in the mouth, tongue, and face (Layragues, 2001). Other associated neurological symptoms may include disorder of speech – slow, slurred, and monotonous – which is a constant finding.

Impairment of intellectual function, cerebellar ataxia, and tremor may also be encountered, as may be hypoerreflexia and extrapyramidal rigidity. Episodes of confusion with or without precipitating factors may be part of the clinical pictures but they are variable in frequency and severity (Layragues, 2001).

The most important clinical aspect is the precocious treatment of cirrhotic patients with movement disorders because neurological symptoms often become irreversible and the success of the treatment depends on the degree of neuronal loss observed in these patients. Medical treatment with ammonia lowering agents – lactulose, metronidazole, neomicine and sodio benzoate – can be tried first but their beneficial effects are variable. Usefulness of dopaminergic agonists such as levodopa (Layragues, 2001) or bromocriptine (Layragues, 2001) in general is controversial and the specific effects of these drugs on movement disorders associated with hepatic encephalopathy have never been tested adequately. Liver transplantation should be considered if patients fit the eligibility criteria (Layragues, 2001).

In: Basal Ganglia and Thalamus
Editors: R. Moretti, P. Torre and R. M. Antonello
ISBN 978-1-60692-198-2


*Chapter 12*

# Basal Ganglia and Language

***Rita Moretti, Paola Torre and Rodolfo M. Antonello***

Cerebral representations of language include cortical and subcortical processes. However, research on the cerebral cortex has overshadowed that on subcortical functions (Petrovici, 1980). It is now widely believed that neuropsychological functions are not simply localized in specific cortical areas, but rather are subserved by a mosaic of neuronal structures comprising cortical regions, subcortical structures and their respective reciprocal connections (Damasio, 1994; McCarthy and Warrington, 1990; Shallice, 1988 ). Language related cortical zones are richly interconnected with other cortical regions (particularly via myelinated fibre tracts), but they also participate in subcortical feedback loops within the basal ganglia (caudate and putamen) and the thalamus (Goldman-Rakic and Selemon, 1990). In fact, through their widely diffuse cortical connections subcortical structures regulate higher functions such as language processing, memory, writing etc., traditionally considered cortical-dependent (Signoret, et al. 1984; Alexander et al., 1987). Evidence accumulated to date has highlighted the complexity of the subcortical aphasia problem. For example, reduced subcortical activation may follow cortical lesions that produce aphasia (PET) (Metter et al.,1988). Given such methodological conundrums, it has been difficult to explain the contribution of each of the subcortical circuits to language processing (Gurd and Bamford, 1997). Lesions to the dominant basal ganglia (caudate and putamen) and thalamus do not invariably give rise to aphasia. Stereotactic neurosurgical operations (Ojemann, 1982) and the development of increasingly sophisticated neuroimaging techniques such as computerized tomography (CT) and Magnetic Resonance Imaging (MRI) (Damasio and Damasio, 1989) have shown that specific subcortical structures of the dominant hemisphere and, in particular, the basal ganglia, may play a fundamental role in language processing. Further studies (Crosson, 1992) have shown that the cortico-striato-thalamo-cortical loop is also involved in language processing. Stimulation to the head of the left caudate has been associated with reduction in sentence length and semantic coherence, whereas stimulation to the white matter anterior to the head of the caudate has only been associated with arrest on ongoing speech (Crosson, 1992). Unfortunately, the extent of spreading activation is difficult to assess and the relationship between the performance of neurological patients and normal subjects is unclear in such studies. The head of the caudate projects to Broca's area as well as to prefrontal areas via pallido-nigral and thalamic relays (Rolls and Johnstone, 1992; Saint-Cyr et al., 1992). One might expect then, that lesions to the head of the left caudate nucleus would be accompanied by language impairment in the lesion phase (Crosson, 1992); the acute phase lasts up to 3 months post-stroke, and recovery may be rapid: anatomical correlates are difficult to specifiy during this phase as behavioural impairments may be due to remote effects of the lesion. During the lesion phase, 3-4 months post-onset, the lesion itself has the main effect on the clinical presentation. In the late phase, the clinical presentation shows the effects of the recovery process, in addition to those of the resolving lesion (Alexander, 1989). Few clinical studies concerning subcortical lesions have shown typical language disorders (Signoret, et al., 1984; Alexander et al., 1987; Fabbro, 1997). It has been recognized recently that the first case of aphasia described by Broca, along with the well-known lesion of the foot of the third circumvolution of the frontal cortex, had a very large subcortical lesion (Fabbro, et al., 1997; Signoret et al.,1984). In fact, lesions in the left thalamus can determine a reduction of spontaneous speech, mild impairment of articulation, paraphasias and perseverations (Mohr et al., 1975). Lesions of the left basal ganglia (with or without the involvement of the adjacent white matter) provoke abnormalities in the speech and language output systems, characterized by disturbances in speech rate and volume, and by transcortical motor aphasia (Aglioti and Fabbro, 1993; Fabbro et al., 1996). The basal ganglia are also involved in the retrieval of information from the storage of nondeclarative (implicit) memory (Squire and Knowlton, 2000). This type of memory may be anatomically organized in cortical and subcortical structures.

Relative to linguistic processes, contemporary theories of subcortical participation in language have endorsed a role for the globus pallidus internus (GPi) in the control of lexical–semantic operations.

However, attempts to empirically validate these postulates have been largely limited to neuropsychological investigations of verbal fluency abilities subsequent to pallidotomy.

Whelan et al. (2004) evaluated the impact of bilateral posteroventral pallidotomy (BPVP) on language function across a range of general and high-level linguistic abilities, and validated/extended working theories of pallidal participation in language. Comprehensive linguistic profiles were compiled up to 1 month before and 3 months after BPVP in 6 subjects with Parkinson's disease. Commensurate linguistic profiles were also gathered over a 3-month period for a non-surgical control cohort of 16 subjects with PD and a group of 16 non-neurologically impaired controls. Nonparametric between-groups comparisons were conducted and reliable change indices calculated, relative to baseline/3-month follow-up difference scores. Group-wise statistical comparisons between the three groups failed to reveal significant postoperative changes in language performance (Whelan et al., 2004).

Case-by-case data analysis relative to clinically consequential change indices revealed reliable alterations in performance across several language variables as a consequence of BPVP. These findings lend support to models of subcortical participation in language, which promote a role for the GPi in lexical– semantic manipulation mechanisms. Concomitant improvements and decrements in postoperative performance were interpreted within the context of additive and subtractive post-lesional effects. Specific declines in the ability to retrieve lexical items defined by their affiliated motor programs, subsequent to BPVP, may suggest potential ramifications for our current conceptualization of how semantic information is neurally coded. Frontal lobe processes have been implicated in the planning and execution of movement (Whelan et al., 2004). On the basis of this evidence, the frontal lobes have also been affiliated with the potential storage of semantic codes relevant to an object's sensory–motor features. Subsequently, lesions within components of the basal ganglia associated with the regulation of cortically driven movements, may plausibly evoke declines in tool-cued fluency performance (Whelan et al., 2004). Extending the tenets of the Lexical Decision Making model, the flexible exploration and manipulation of lexical elements potentially generated within the frontal as well as temporo-parietal cortex, may be implicated after lesions of the GPi (Whelan et al., 2004).

Results for basal ganglia seem far from the conclusive final set point.

Longoworth et al. (2005) tested whether striatal damage or loss of its dopaminergic innervation is reliably associated with grammatical rules which combine procedural memory and morpheme combination (selective regular past tense deficits) in patients with subcortical cerebrovascular damage, Parkinson's disease or Huntington's disease. Longsworth et al. (2005) findings do not support an essential role for the basal ganglia in applying grammatical rules to combine morphemes in language production or to decompose words into morphemes in language comprehension. This is consistent with a previous report that patients with Parkinson's disease did not show a significant difference between untimed past tense elicitation for regular and irregular verbs, matched for frequency and were faster and less error-prone in speeded regular past tense production than irregular (Almor et al., 2002). Several aspects of Longsworth et al. (2005) results are compatible, however, with the hypothesis that the striatum plays a restricted, non-language specific, inhibitory role in the late stages of language processing. First, the automatic activation of semantic and morphological representations in language comprehension was intact in all groups as discussed above.

Secondly, patients with moderate Parkinson's disease showed an abnormal pattern of response facilitation for phonologically related words in language comprehension.

This suggests that despite normal activation of lexical representations initially consistent with incoming speech, there is a failure to suppress activation for those representations that subsequently become incompatible with the speech signal. Thirdly, patients across all groups examined showed evidence of an inability to suppress semantically appropriate alternatives when trying to inflect novel verbs and this was the dominant error made by patients with moderate Huntington's disease across all conditions of both tasks. This appears to be consistent with reports of semantic paraphasias following subcortical cerebrovascular damage (Damasio et al., 1982) and the failure of patients with striatal dysfunction to suppress the infrequent meaning of homophones in semantic priming at long stimulus onset intervals (Copland, 2003).

Longsworth et al. (2005) results add to the increasing evidence that the role of the striatum in the language system is restricted to late integrational processes requiring inhibition of competing alternatives.

Numerous aspects of language have been shown to be affected after basal ganglia selective damage, with no clear relationship between the presence and type of linguistic deficits in relation to the site of injury (Nadeau and Crosson, 1997; Kertesz, 1992; Weiller et al., 1990, 1993). Moreover, infarctions involving the left basal ganglia do not always lead to classical aphasic syndromes (Nadeau and Crosson, 1997; Cappa et al., 1983; Fromm et al., 1985; Basso et al., 1987; Wallesch et al., 1997) or to selective deficits in aspects of language typically associated with left-hemisphere function (Rowan et al., 2007).

In seeking to explain discrepancies in the literature on brain structure–function relationships, the possibility arises that in some cases there may have been brain damage that was not detected using the imaging methods that were available at the time. In such cases, erroneous conclusions may well have been reached about the structural basis of functional abnormalities. To take the example of patients with infarctions of the basal ganglia, it may be that in some cases there is additional, more subtle pathology that is not seen on conventional MRI, possibly in connected fiber pathways and cortical sites (Werring et al., 2000). Perhaps it is the pathology in these sites that is primarily responsible for any language impairments that are seen. In line with this, Hillis et al. (2002) showed that aphasia or neglect in 44 patients with acute subcortical infarctions was consistently related (in 100% of cases) to cortical hypoperfusion in the middle cerebral artery (MCA) territory. The same group went on to show that, in patients with acute infarction of the left caudate nucleus, the presence and type of aphasia could be explained by the site and extent of cortical hypoperfusion (Hillis et al., 2004). In addition to perfusion imaging, a number of additional MRI approaches offer opportunities for identifying subtle brain abnormalities that are not seen on conventional neuroradiological assessment (Rowan et al., 2007). These include voxel-based morphometry (VBM) analysis of structural T1-weighted MRI data sets (Ashburner and Friston, 2000), which can be used to identify subtle gray or white matter abnormalities and their relationships to cognitive function (Isaacs et al., 2001; Watkins et al., 2002). VBM can also be used, inconjunction with diffusion tensor imaging (DTI) (Basser and Pierpaoli, 1996), to identify abnormalities in fiber tracts, via changes in diffusion anisotropy (Klingberg et al., 2000; Eriksson et al., 2001).

A group of children and young adults who had suffered arterial ischemic stroke resulting from a single unilateral infarction apparently confined to the basal ganglia and surrounding white matter were selected by Rowan et al. (2007). The selection was made on the basis of conventional neuroradiological examination of T1- and T2-weighted MRI scans carried out for clinical purposes after the stroke. All patients had lesions that were acquired during infancy or childhood, with onset after the age of 6 months. Exclusion criteria were as follows: (i) involvement of thalamic or cortical regions on clinical MRI; (ii) hemorrhage on clinical MRI; (iii) sickle cell disease; (iv) moyamoya syndrome; (v) cardiac abnormalities; (vi) unresolved epilepsy; and (vii) pre-injury neurological difficulties, developmental delay or difficulties in speech and language, according to parental questionnaires (Rowan et al., 2007). All patients were required to be at least 7 years old at the time of the MRI scans for this study, so that they could undergo scans without sedation, and were required to be native speakers of English (Rowan et al., 2007).

All patients showed single unilateral lesions on visual inspection of conventional T1- and T2-weighted MRI scans. These lesions were confined to the basal ganglia and surrounding white matter, including extension into the external and extreme capsules and the corona radiata in some cases. Neuro-radiological assessment showed no obvious relationship between extent or site of basal ganglia damage and language performance (Rowan et al., 2007). For example, no common site of damage was observed in the basal ganglia in relation to language performance in the left-hemispheric injured group. More specifically, the head of the left-hemisphere caudate nucleus was not damaged in all patients with language deficits and was damaged in some patients who showed no evidence for language deficits (Rowan et al., 2007). Hemodynamic abnormalities distant from the core lesion site were observed in seven patients, all on the side of the infarction. The three patients with left-hemisphere injuries and the poorest language function evidence of left cortical hemodynamic abnormalities that involved fronto-parietal regions and included cortical language areas in all three cases. No other patients with left or right-hemisphere injuries showed any evidence of hemodynamic abnormalities in the left cortical language areas. All of the patients who had hemodynamic abnormalities on PWI also showed Magnet angiography (MRA) abnormalities that involved residual stenosis of the middle cerebral artery (MCA) or internal carotid artery, or occlusion of the MCA. Rowan et al. (2007) observations are consistent with a number of previous studies, which have indicated that language difficulties after basal ganglia infarctions in adults are related to abnormalities remote from the core of the lesion, including white matter tract damage (Alexander et al., 1987; Naeser et al., 1982), cortical disconnection (Perani et al., 1987), and cortical hemodynamic abnormalities (Hillis et al., 2002, 2004). The causal nature of the relationships among the cortical gray matter loss, white matter changes, cortical hemodynamic abnormalities, the basal ganglia lesions, and the manifestation of language impairments remains unclear, especially as the damage in these patients occurred relatively early in life and might therefore have led to an abnormal developmental trajectory. Damage to the basal ganglia, or to white matter surrounding the basal ganglia, is thought to result in the anterograde degeneration of connected fiber pathways and connected cortical sites that are remote from the core site of the lesion (Werring et al., 2000). The possibility that functional deactivation in the cortex leads to degeneration of fibers projecting from the cortex to the basal ganglia, however, cannot be excluded on the basis of these results. Cortico-thalamic disconnection may also play a role, as discussed by Nadeau and Crosson (1997). Another possibility is that blood flow abnormalities in cortical language areas cause language deficits and a reduction in functional activity leads to degeneration of gray matter in these sites. In support of this possibility, Weiller et al. (1990) observed regions of focal cortical atrophy on MRI one year after infarction that were not visible acutely in regions that were previously hypoperfused. The converse situation for the hemodynamic abnormalities, in which degeneration of cortical gray matter resulting from a lack of basal ganglia input and/or white matter tract damage leads to a reduction in recruitment of blood to these regions, could also be true (Perani et al., 1987). A combination of these factors might explain language impairments following basal ganglia infarctions, and the influence of these different mechanisms might vary between individuals (Rowan et al., 2007).

More recently, speculative clinical lesional systems have been substituted by functional studies. Functional magnetic resonance imaging data (fMRI) was used to determine the frontal, basal ganglia and thalamic structures engaged by lexical status of generated items, the use of semantic versus phonological information during language generation, and rate of generation (Crosson et al., 2003). Components of a left pre supplementary motor area (SMA)-dorsal caudate nucleus-ventral anterior thalamic loop were active during word generation from rhyming or category cues, but not during nonsense syllable generation. Findings indicate that this loop is involved in retrieving words from pre-existing lexical stores. Relatively diffuse activity in the right basal ganglia (caudate and putamen) also was found during word-generation tasks, but not during nonsense syllable generation; therefore, Crosson et al. (2003) suggested that the right basal ganglia activity serves to suppress right frontal activity, preventing right frontal structures from interfering with language production and control.

Booth et al. (2007) looked at effective connectivity (the influence that one brain region has on another) of the cerebellum and basal ganglia with regions thought to be involved in phonological processing, i.e. left inferior frontal gyrus and left lateral temporal cortex. The results obtained by fMRI analysis showed that the cerebellum has reciprocal connections with both left inferior frontal gyrus and left lateral temporal cortex, whereas the putamen has unidirectional connections into these two brain regions. Furthermore, the connections between cerebellum and these phonological processing areas were stronger than the connections between putamen and these areas (Booth et al., 2007). This pattern of results suggests that the putamen and cerebellum may have distinct roles in language processing. Based on research in the motor planning and control literature, the authors argue that the putamen engages in cortical initiation while the cerebellum amplifies and refines this signal to facilitate correct decision making (Booth et al., 2007).

Xiao et al. (2005) used a fast event-related functional magnetic resonance imaging (fMRI) design and contrasted the lexical processing of pseudowords and real words. Participants carried out an auditory lexical decision task on a list of randomly intermixed real and pseudo Chinese two-character (or two-syllable) words. The pseudowords were constructed by recombining constituent characters of the real words to control for sublexical code properties. pseudowords and the real words both activated bilateral auditory cortex in the superior temporal gyrus (BA21/22), as one would expect for the present task with auditory stimulation. Activation in supplementary motor cortex, precentral sulcus, cingulate gyrus, and subcortical structures such as cerebellum and caudate mostly likely reflects the planning and execution of motor responses, which like the stimulus perception process merged with the lexical decision process in their fMRI responses. Thalamus activation may reflect the involvement of attentional processes (Xiao et al., 2005). The remaining activations at middle temporal gyrus, inferior frontal gyrus, and supramarginal gyrus were all located in language-related areas that have been well documented in imaging studies of linguistic processes (see Data and Literature in Xiao et al., 2005).

Overall, the pattern of activation was highly comparable for the two types of stimulus. All regions activated in the pseudoword condition except the parahippocampal gyrus were also activated in the real word condition. Most of these overlapping regions did not show up in the direct contrast between the two conditions, indicating a close match in loci, activation intensity, and spatial extent of corresponding regions (Xiao et al., 2005).

Basal ganglia have been studied even in word retrieval in old healthy volunteers (Wierenga et al., 2008). Word retrieval requires access to memories for word forms (i.e., lexical forms) and meaning (i.e., semantics). Word finding difficulties in healthy aging may result either from a deterioration of frontal neural substrates that underlie the ability to strategically access and retrieve appropriate lexical word forms or from a deterioration of the semantic stores that represent meaning, thought to reside, at least in part, in the inferior temporal lobe (Ishai et al., 1999). Frontally mediated language operations involved in word retrieval may include searching semantic stores, monitoring and evaluating potential words to represent a concept, and exerting cognitive control to select the appropriate word from among competing alternatives (Carter et al., 2000). Prior studies have found that medial frontal structures, especially those at the border of pre-SMA and the rostral cingulated zone, are involved in initiation of language, cognitive control and monitoring conflict between competing responses (Damasio and Anderson, 1993; Crosson et al., 1999; Carter et al., 2000; Wierenga et al., 2008).

The lateral frontal cortex of the language dominant hemisphere, including Broca's area, cortex along the inferior frontal sulcus, and possibly pars orbitalis, is involved in selection, retrieval and execution of the lexical–semantic response appropriate to internal and external constraints (Damasio and Anderson, 1993; Gabrieli et al., 1998; Crosson et al., 1999; MacDonald et al., 2000; Wagner et al., 2001). Given the neural circuitry between the basal ganglia and the frontal lobes and evidence that regions of the basal ganglia are involved in language production tasks, age-related changes may also be evident in subcortical regions (Alexander et al., 1986; Crosson et al., 2003; Wierenga et al., 2008). On the other hand, although semantic knowledge is thought to be widely distributed throughout the brain, there are relatively consistent findings implicating the fusiform gyrus in semantic processing of category domain and visual attributes (Ishai et al., 1999; Wierenga, 2005). Twenty younger and 20 older healthy adults named aloud photographs during event-related fMRI (Wierenga et al., 2008). A major finding of Wierenga et al. (2008) study is that older adults demonstrated a larger frontal network of activity during word retrieval than younger adults. When compared directly to younger adults, not only do older adults show greater activity in left-hemisphere regions traditionally associated with language (i.e., rostral cingulate zone, bilaterally and SMA), but they demonstrate less lateralization of activity as evidenced by an increase in right frontal activity. In particular, older adults show greater activity in Broca's area homologue (BA 45) in the right hemisphere, an anterior region of the right inferior frontal gyrus, and the right anterior cingulated (Wierenga et al., 2008).

This increase in activity occurred in the context of equivalent naming accuracy. The equivalent accuracy rates between older and younger adults suggest that there is no gross compromise in semantic substrates. The significantly slower retrieval of words for the older subjects suggests that they may have some compromise in word retrieval. These results suggest that word-finding problems are likely due to difficulty accessing lexical–semantic information rather than a deterioration of information in the neural substrates for semantic memory (Wierenga et al., 2008).

Despite no significant difference in overall naming accuracy between groups, when individual performance (e.g., naming accuracy) was taken into account, increased BOLD response in the right inferior frontal gyrus (e.g., pars orbitalis) for older adults was positively correlated with greater naming accuracy, suggesting that at least for some older adults recruitment of right-hemisphere homologues is in fact beneficial for performance (Wierenga et al., 2008). However, a positive correlation between naming accuracy and BOLD response was not found uniformly across regions of the right hemisphere. In contrast, the BOLD response was found to negatively correlate with naming accuracy in older adults in the right precentral gyrus (Wierenga et al., 2008).

Taken together, contrary to the idea that additional activity is universally compensatory, the findings of this work (Wierenga et al., 2008) suggest that while activity in some right frontal regions is associated with enhanced performance, activity in other right-hemisphere regions is not. This regional specificity within the right hemisphere is consistent with Naeser's recent findings in aphasia in which suppression of right pars opercularis via repetitive transcranial magnetic stimulation made language impairments more severe, whereas suppression of right pars triangularis slightly improved picture naming (Naeser et al., 2005). Taken together, these results suggest that the ability to recruit contralateral regions of the inferior frontal gyrus (in the right hemisphere) including Broca's area homologue and pars orbitalis assists in performance accuracy in older adults but is not beneficial in younger adults (Wierenga et al., 2008).

Activity in the basal ganglia was primarily lateralized to the left hemisphere and was associated with both measures of performance (e.g., accuracy, response latency) in older adults collapsed across performance groups. The left medial globus pallidus and right thalamus and subthalamic nucleus were associated with increased naming accuracy, whereas the left lentiform nucleus and thalamus were associated with faster response time(Wierenga et al., 2008). A previous work in young adults has implicated the dominant pre-SMA-dorsal caudate-ventral anterior thalamic loop in biasing selection toward a particular pre-existing lexical representation during word generation when multiple choices are possible (Crosson et al., 2003). This is consistent with conceptualizations of the basal ganglia as facilitating desired and suppressing undesired behaviours (Mink, 1996; Wierenga et al., 2008).

In: Basal Ganglia and Thalamus
Editors: R. Moretti, P. Torre and R. M. Antonello ISBN 978-1-60692-198-2


*Chapter 13*

# Basal Ganglia and Polylinguism

***Rita Moretti, Paola Torre, Rodolfo M. Antonello, Antonio Bava****
* Dip. Fisiologia e Patologia, Università di Trieste, Italy

Stereotactic neurosurgical operations (Ojemann, 1982) and the development of increasingly sophisticated neuroimaging techniques such as computerized tomography (CT) and Magnetic Resonance Imaging (MRI) (Damasio and Damasio, 1989) have shown that specific subcortical structures of the dominant hemisphere and, in particular, the basal ganglia, may play a fundamental role in language processing. Further studies (Crosson, 1992; Crosson et al., 2003; Wierenga et al., 2008) have shown that the cortico-striato-thalamo-cortical loop is also involved in language processing. Stimulation to the head of the left caudate has been associated with reduction in sentence length and semantic coherence, whereas stimulation to the white matter anterior to the head of the caudate has only been associated with arrest on ongoing speech (Crosson, 1992). Unfortunately, the extent of spreading activation is difficult to assess and the relationship between the performance of neurological patients and normal subjects is unclear in such studies. The head of the caudate projects to Broca's area as well as to prefrontal areas via pallido-nigral and thalamic relays (Rolls and Johnstone, 1992; Saint-Cyr et al., 1992).

One might expect then, that lesions to the head of the left caudate nucleus would be accompanied by language impairment in the lesion phase (Crosson, 1992; Crosson et al., 2003; Wierenga et al., 2008); the acute phase lasts up to 3 months post-stroke, and recovery may be rapid: anatomical correlates are difficult to specifiy during this phase as behavioural impairments may be due to remote effects of the lesion. During the lesion phase, 3-4 months post-onset, the lesion itself has the main effect on the clinical presentation. In the late phase, the clinical presentation shows the effects of the recovery process, in addition to those of the resolving lesion (Alexander, 1989; see the previous chapter for recent data on the topic).

Few clinical studies concerning subcortical lesions have shown typical language disorders (Signoret, et al., 1984; Alexander et al., 1987; Fabbro, 1997; Crosson et al., 2003; Wierenga et al., 2008). It has been recognized recently that the first case of aphasia described by Broca, along with the well-known lesion of the foot of the third circumvolution of the frontal cortex, had a very large subcortical lesion (Fabbro et al., 1997; Signoret et al.,1984).

In fact, lesions in the left thalamus can determine a reduction of spontaneous speech, mild impairment of articulation, paraphasias and perseverations (Mohr et al., 1975). Lesions of the left basal ganglia (with or without the involvement of the adjacent white matter) provoke abnormalities in the speech and language output systems, characterized by disturbances in speech rate and volume, and by transcortical motor aphasia (Aglioti and Fabbro, 1993; Fabbro, et al., 1996).

The basal ganglia are also involved in the retrieval of information from the storage of non-declarative (implicit) memory (Squire and Knowlton, 2000; Longworth et al., 2005). This type of memory may be anatomically organized in cortical and subcortical structures. On the other hand, declarative (explicit) memory is subserved by the fronto-temporal cortical regions and by the hippocampus and amygdala (Squire and Knowlton, 2000; Longworth et al., 2005).

We described (Moretti et al., 2003 -C) a patient who developed an impairment of the mother tongue after an infarct of the caudate. During follow-up, a dramatic improvement of the mother tongue accompanied by worsening of the second language became evident after the extension of the ischemic lesion to the cortex. This case provides evidence not only of the double dissociation between the two languages spoken by the patient, due to left basal ganglia lesion, but also of the different roles of the subcortical and cortical structures which regulate the implicit and explicit memory and organization of the strategies for language retention (Nadeau and Crosson, 1997).

A hypertensive 46-year old, right handed woman (+24 test of Briggs and Nebes, 1975) abruptly presented on July, 14$^{th}$ 1997 with a mild right motor hemisyndrome, which evolved over a few hours in a severe right hemiplegia associated with a period of mutism lasting about ten days. When examined after two weeks, at the end of July, 1997, she was fully alert, cooperative and oriented to place and time. Cerebral CT on admission was normal. Carotid duplex ultrasonogram showed a patent right internal carotid artery and a suspected occlusion of the left internal carotid artery. Carotid artery angiography confirmed severe stenosis (more than 90%) of the left internal carotid artery, consistent with a dissection of the artery. Two weeks later a cerebral CT showed a hypodense lesion of 1.5 cm involving the left caudate nucleus. At this time, the patient began to speak again and her linguistic ability was investigated. One month after the admission she underwent a second carotid duplex ultrasonogram which showed patency of both the right and the left internal carotid artery. The patient entered a neuromotor rehabilitation program without any formal language re-education. Four months after the stroke, in November 1997, she was discharged as functionally independent, though showing distal weakness in the right arm and clumsiness of the right hand. During follow-up, the patient suffered five episodes of angina pectoris. The consequent low cardiac output determined an extension of the ischemic area to frontal cortical region indicated by a 99m-Tc-ECD scintigram, which put in evidence mild hypoperfusion of the left fronto-temporal cortex and by a MRI, which indicated a diffuse ischemic lesion in the left fronto-temporal hemisphere. A second neurolinguistic examination was performed five months after the stroke, in December, 1997 (Moretti et al., 2003- C)

The patient was a Croatian (L1) woman who had spoken Croatian since she was born. She attended primary and secondary school in Croatia. She immigrated to Italy at 21 years of age and she learned Italian (L2) working in Trieste. She married an Italian man, and she spoke Italian since then. She watched Italian television, read Italian magazines, and attended for two years formal courses in Italian. She spoke Croatian every second day, when she phoned relatives in Croatia and when she met Croatian friends in Trieste. During the two months after the stroke, the patient's mother tongue was so severely impaired that she could hardly interact linguistically with her family and friends using L1. She noted that speaking Italian to her relatives and to the nurses or the doctors was far easier than speaking Croatian. Afterwards, she realized that L1 had improved, whereas L2 had seriously worsened.

During the first examination, we collected a five-minute sample of spontaneous speech extracted from 15 minutes in both L1 and L2, recorded on tape. She was tested by a bilingual (Croatian-Italian), who was previously trained in the administration of the Bilingual Aphasia Test. We then tested her capacity to retrieve a short story in L1 and L2 and her denomination capability (Paradis, 1987; Paradis and Canzanella, 1990). For this purpose we chose parts of the Bilingual Aphasia Test (BAT), standardized for age and education (Paradis, 1987; Paradis and Canzanella, 1990). We also tested her digit-span forward and backward (Wechsler, 1945), her mental arithmetic performance and her reading capabilities (Paradis, 1987; Paradis and Canzanella, 1990) in L1 and L2. Her spontaneous language underwent neurolinguistic analysis. Four months later she performed a complete BAT test (Paradis, 1987; Paradis and Canzanella, 1990) in both languages, the Mini-Mental State Examination (Folstein et al., 1975), the Hamilton Depression Scale (Hamilton, 1960) and to the Raven Standard Progressive Matrices (Raven, 1960). A Fisher's exact test and chi squared test were employed to analyze the data obtained in the two sessions. As far as spontaneous language of the patient was concerned, all the data have been reported in table 1 (where we include data regarding spontaneous language parameters of a healthy group of volunteers).

**Table 1. A synopsis of the linguistic analysis of the subject in the two languages and in the two sessions**

| Measures | L1 | | L2 | | Statistical parameters | Normal Range |
|---|---|---|---|---|---|---|
| | July | December | July | December | | |
| Words/minute | 68 | 70 | 55 | 89 | Ns | 92 |
| Total number of words | 359 | 347 | 279 | 445 | Ns | 478 |
| utterances | 18 | 20 | 13 | 25 | Ns | 32 |
| Verbal paraphasias | 3 | 0 | 1 | 1 | Ns | 0 |
| Substitutions of free morphemes | 9 | 3 | 4 | 6 | Ns | 1 |
| Semantic paraphasias | 6 | 0 | 0 | 2 | $p < 0.05$ | 1 |

**Table 1. (Continued)**

| Measures | L1 | | L2 | | Statistical parameters | Normal Range |
|---|---|---|---|---|---|---|
| | July | December | July | December | | |
| Omission of free grammatical morphemes | 3 | 2 | 0 | 9 | $p < 0.05$ | 0 |
| Semantically deviant sentences | 4 | 0 | 0 | 4 | $p < 0.05$ | 0 |
| Synctatic comprehensio n tasks (mistakes) | / | 7 | / | 16 | $p < 0.00001$ | 11 |
| Morphological tasks (mistakes) | / | 4 | / | 9 | $p < 0.001$ | 4 |
| Denomination (mistakes) | 2 | 1 | 2 | 3 | Ns | 0 |
| Retrieval of a story (mistakes) | 4 | 1 | 2 | 0 | Ns | 1 |
| Reading task (mistakes) | 1 | 6 | 0 | 9 | Ns | 2 |
| Word fluency | / | 14 | / | 4 | $p < 0.001$ | 32 |
| Mental arithmetic (mistakes) | / | 8 | / | 9 | Ns | 2 |
| Writing words | / | 14 | / | 12 | Ns | 123 |
| Writing utterances | / | 6 | / | 4 | Ns | 23 |
| Writing types | / | 7 | / | 7 | Ns | 56 |
| Phonological paragraphias | / | 2 | / | 5 | $p < 0.05$ | 1 |
| Substitutions of free grammatical morphemes | / | 3 | / | 2 | Ns | 0 |
| Semantically deviant sentences | / | 0 | / | 4 | $p < 0.05$ | 0 |

Her digit span forward was 6 numbers; her digit span backward was 5 numbers (Wechsler, 1945). Between the first and the second examination, the linguistic situation remained more or less the same. She was exposed to L2, speaking to her husband and to the nurses who followed her. She talked to her relatives and her friends for one hour daily in L1. Thus, the exposure to Croatian and Italian during that period was similar to pre-stroke conditions. The same bilingual tested her five months later (December, 1997) and the patient performed a more complete battery of tests. She was not depressed according to the results of the Hamilton Depression Scale (Hamilton, 1960). She obtained a full scale IQ of 105 on the Raven Standard Progressive Matrices (Raven, 1960), which represents a result within the $71^{st}$ percentile, standardized for age and education. She was more at ease speaking L1 than L2, in that she could produce only short sentences with low voices and great effort in L2 (data have been summarized in Table 1).

Statistical analysis indicated that there were no statistical differences (Fisher's exact test) between the number of types, verbal paraphasias, substitutions of free grammatical morphemes, the number of mistakes in the retrieval of a story, in naming, in calculation and in reading tasks which the patient produced in July and in December. In contrast, there were significant differences between the number of semantic paraphasias produced in the two languages in July and in December (6 in L1 and 0 in L2 in July; 0 in L1 and 2 in L2 in December; $p<0.05$). The number of omissions of free grammatical morphemes and of full verbs made in the two languages and in the two trials also varied significantly (3 in L1 and 0 in L2 in July; 2 in L1 and 9 in L2 in December; $p<0.05$). Semantically deviant sentences also differed significantly (4 in L1 and 0 in L2 in July; 0 in L1 and 4 in L2 in December; $p<0.05$).Word fluency differed significantly (14 words in L1 and 4 in L2 in December, normal range= 32, $p<0.001$), as well as the difference in syntactic comprehension tasks in December (7 mistakes in L1, 16 in L2 ; normal error range = 11, $\chi^2$ test $p<0.00001$) and the difference between the number of mistakes produced in the morphological tasks ( 4 mistakes in L1, 9 in L2; normal error range = 4, $\chi^2$ test $p<0.001$. Finally, the number of phonological paragraphias (2 in L1 and 5 in L2 in December, normal range = 1, $p< 0.05$) and the number were also significant (Moretti et al, 2003- C).

Many studies suggest the involvement of subcortical structures in language production. Both the caudate and pallidal stimulation resulted in the arrest of ongoing speech (Van Buren, 1963; Ojemann, 1982); in particular, it was noted that the stimulated patient seemed unaware that anything out of ordinary had happened. Crosson (1992) proposed that the basal ganglia regulate the release of cortically generated language segments, and Wallesh and Papagno (1988) suggested that the basal ganglia participate in the selection of the most appropriate lexical alternatives, originally generated by cortical regions. The head of the caudate receives extensive (and exclusive) projections from the dorso-lateral refrontal cortex and other association cortices of the frontal temporal and parietal lobes. Language deficits following subcortical lesions may, in part, be due to overlying cortical vascular effects of the lesion with perfusion sufficient to maintain cell viability but insufficient for normal function (Vallar et al., 1988).

Again, language deficits may also result from cortical deafferentation; this in itself does not preclude the logical possibility that particular regions of the basal ganglia and thalamus may play specific roles in speech and language (Olsen et al., 1984).

Olsen et al. (1984) claimed that behavioural symptomatology of the "subcortical aphasias" is due to the vascular effects of overlying cortical hypoperfusion rather than to any direct functional effects of the subcortical lesions themselves.

However, there were several weaknesses in their study (Steinmetz and Seitz, 1997) and the CT scans were of low resolution. It is also possible, by their own admission, that the non aphasic patients also had reduced cortical blood flow relative to normal control subjects. The claim for a direct role for subcortical structures in language receives strong support from Metter et al (1988): their evidence was derived from CT and PET scan studies of patients with left-hemisphere cerebral infarcts or haemorrhages. The most interesting finding is that fluency deficits were associated with lesion to the basal ganglia, thalamus and/or internal capsule, as demonstrated by Gurd et al. (1997): nonetheless, as demonstrated in another study (Gurd et al., 1988). It is possible to suffer an isolated infarct of the anterior limb of the internal capsule with speech, but no language, deficits. More recent studies concluded that vascular lesions of the dominant hemisphere involving large portions of the head of the caudate nucleus and/or the putamen usually result in aphasia, although exceptions occur (Wallesch, 1997). According to Weiller et al. (1993) the type and severity of aphasia vary widely, and there is little or no correlation between localization or type of infarct and type of mistakes found (data confirmed by Kumral et al., 1999). Further evidence may come from the study of Parkinson's disease (PD); in the PD model, the basal ganglia-thalamo-cortical language circuit is not lesioned, but nonetheless does not function normally as an integrated system: this provides yet another window on the brain-language functional interface, which takes into account the functions of the neural linguistics sysytem as a whole (Gurd, 1995). The selectively reduced cortical hypometabolism in PD is not due to a vascular effects in the sense in which they are implicated in the subcortical aphasias, however, it is a good reason to look for language deficits, in particular of word-finding in this population of patients with progressive bilateral dysfunction of "subcortical-cortical circuitry" (Moretti et al., 2003-A-B).

Preliminary functional magnetic resonance imaging (f- MRI) studies indicate a role for the caudate nucleus in the generative aspects of language (Crosson et al., 1997). From a clinical point of view, caudate lesions can determine nearly every possible degree of impairment in the various aspects of language. Some patients are fluent but have moderate to severe impairment in comprehension, repetition and naming; some have nonfluent output, with or without impairment of comprehension. Others have no language disorder at all (Cappa et al., 1983; Alexander et al., 1987; Nadeau and Crosson, 1997).

The prefrontal association cortex projects massively to the caudate nucleus and very little to the putamen, in an orderly fashion; projections from the orbitofrontal and anterior cingulated cortex target the striosmes, those from the lateral prefrontal convexity target the matrisomes, and the rest (from the limbic area such as amygdala) target the matrix (Cummings, 1995). Central to the functional spectrum of the caudate nuclei are the so-called tonically active neurons (TANs), specialized cells that are set to change their firing patterns in response to stimuli that are generally perceived to have behavioural significance (e.g.:

rewarding stimuli): placed at the matrisome-striosome border junction, they may help to select and generate new striatal activity patterns in response to the integration of behaviourally significant information. The caudate head would act as a filter or a switch among parallel circuits originating from the frontal cortex. The caudate heads may integrate the attentional axis (anterior cingulated) and the planning and behavioural axis (dorsolateral cortex). The dorsolateral prefrontal cortex is crucial for integrating information from all the three main frontal association circuits (dorsolateral/orbitofrontal/mmedial frontal) that can be acted upon through the executive motor circuitry (Canavero, and Fontanella, 1998). Many studies suggest that different subsystems, which include subcortical structures, define different aspects of language production (Alexander, 1989). The system for speech and language initiation includes the supplementary motor area, its efferent projections through the anterolateral periventricular white matter, the dorsolateral region of the frontal lobe and the striatum. The system for speech output involves the frontal operculum and the lower motor cortex, the putamen and the caudate and their connections (Hikosaka, 1991). Acting together, these structures regulate the volume of the spoken output, the fluency and, probably, the initiation of spoken output. There are different points of view regarding whether subcortical damage can produce an aphasia syndrome (Wallesch et al., 1985), which can be summarized by the defective transmission of the so-called "initial thoughts" into language (Luria, 1970). The assumption of functional plasticity of the system would also explain the relative absence of language disturbances in chronic subcortical disease (i.e. Huntington's Chorea, Parkinson's disease, etc.), but also in these cases palilalia, compulsory speech etc. can be found (Matison, Mayeux, Rosen, Fahn, 1982). In our case, the linguistic situation of our patient indicated a more fluent aspect in L1 during the first session and in L2 during the second session (Moretti et al., 2003-C).

As said before, monolingual aphasics who suffer from lesions involving the left basal ganglia present various language deficits (signs of non- fluent aphasia), verbal and semantic paraphasias and even comprehension defects (signs of fluent aphasia) (Moretti et al, 2003-C). The situation is even more complicated in bilinguals or in polyglots with subcortical lesions, where different strategies are involved for learning and speaking different languages. The impairments of speech initiation may show different severities across the two languages, i.e. to be unable to speak the first language, but to be fluent in the second (Aglioti and Fabbro, 1993).

Some patterns of recovery in bilinguals may be best explained by referring to these effects. The most classic case is that described by Gelb (Gelb, 1937) in which a professor of Latin was unable to speak in his mother tongue as a consequence of a lesion presumably located in the left frontal lobe but could speak Latin. Isolated reports (Aglioti and Fabbro, 1993; Aglioti et al., 1996) have described patients with cerebral infarcts confined to the caudate and putamen nuclei of the left hemisphere. These lesions caused severe deficits in the mother tongue, with a good performance in the second language (which the patients rarely spoke before the brain damage). The temporal pattern of ischemic events in our patient included an initial lesion limited to the left caudate nucleus and then, several months later, an hypoperfusion area, which involved the frontal and temporal cortices of the left hemisphere. The neurolinguistic analysis of this bilingual woman may explain the alternative and complementary roles of the cortical and subcortical regions in selecting and speaking different languages, and their importance in the regulation of the implicit and explicit memory and of skilled and automatic functions. In the present study, we found de facto a significantly higher number of omission of full verbs and of free grammatical morphemes in L1 during the first session and in L2 during the second examination, associated to writing disruption in L2. The same was true for the semantically deviant sentences, for the comprehension task and for the morphological tasks. Such findings seem to confirm that deep nuclei may be strictly involved in the "frontal-lobe system" which takes part in language planning and speech production (Aglioti et al., 1996). We observed a partial recovery of the language impairment in L1 due to the caudate nucleus lesion, as in a "transient aphasia" type (Brunner et al., 1982). Then, when the ischemic lesion evolved and the cortical structures were involved, the language disorder consequently changed. A possible hypothesis to explain these aspects is the difference in language acquisition: some authors (Paradis, 1987; Squire and Knowlton, 1995) suggest that the mother tongue, normally acquired early in life, is highly automatized and mainly used through unconscious procedures. In contrast, other languages, generally acquired later in life are typically learned and used through the application of grammatical rules. Different authors (Neville, et al., 1992) proposed that systems for semantic (open class) and grammatical processes (closed class) have a differential maturational time course. Closed class words, which are highly automatic, are mostly represented in the left frontal lobe (see data and Literature in Moretti et al., 2003-C), functionally related to the complex subcortical network involved in automatic motor and cognitive abilities. The subcortical structures are also involved in the retrieval of information from the implicit memory, which is one of the most important mechanism for the basic learning of the mother language (Squire and Knowlton, 2000). For the metalinguistic learning of a second language the process must extend to the explicit memory, which is a cortical domain (Paradis, 1994). The different pattern of acquisition may explain in some cases of neurological impairment a paradoxical recovery with a better performance in the least well-spoken language before the damage. This may be due to the use of a compensatory strategy relying on the metalinguistic knowledge of the second language (Paradis, 1989). The weaker the linguistic competence, the more the speaker will resort to metalinguistic knowledge to control language production. L1 speakers rarely need to rely on their metalinguistic knowledge (Paradis, 1994). There are, in fact, two aspects of lexical items: their acquisition, which is conscious, and their use in a context, which is automatic. Morphosyntax, but not vocabulary, appears to be affected by maturational constraints imposed on procedural memory. Basal ganglia are involved in the implicit memory system and modulate the acquired mechanism (both of cognitive and motor activities) (Squire, 1987; Tulving, 1987; Longsworth et al., 2005). Lesions of these regions result in an evident impairment to the mother tongue. Cortical regions which are involved in explicit memory processes are also involved in the second language, which is acquired with metalinguistic system. Language production and memory tasks may cause an increased vascularization in a large left fronto-temporal region and in the head of the left caudate (Wallesch et al., 1985). The role of the frontal and temporal cortices in our case can be corroborated also by the evidence of a deficit in the syntactic comprehension (traditionally considered as a predominantly hemispheric function) (Fabbro et al., 1997). Moreover, the caudate heads, as already stated, may integrate the attentional axis and the planning axis, which is crucial for integrating information from all three main frontal association circuits to which all the other cortical areas project, directly or via thalamus or basal ganglia relais (Canavero and Fontanella, 1998). Damage in the subcortical structures can both determine a linguistic impairment and a selective disruption of one language (usually the most implicitly learned) in the bilingual patient (Zatorre, 1989).

We have suggested (Moretti et al., 20003-C) that the entire process of language production and comprehension are probably regulated by neural pools whose relay stations are in the cortical and subcortical regions. These neural pools, alone and with their intimate projections, sustain language and writing production. Bilingual cerebral organization may help to understand the role of subcortical regions, intimately bound to implicit memory acknowledgment, and to attentive and to selection/discrimination process. The study of additional bilingual cases will be necessary to understand the fundamental role of a process of learning which will become automatic, in the end and determines the common everyday linguistic output.

In: Basal Ganglia and Thalamus
Editors: R. Moretti, P. Torre and R. M. Antonello
ISBN 978-1-60692-198-2


*Chapter 14*

# Cortical or Subcortical Dementia: A Neuropsychological Comparison

***Rita Moretti, Paola Torre, Valentina Martin*, and Rodolfo M. Antonello***
* Clinica Geriatria, Università degli Studi di Trieste. Italy

Frontal lobe dementia (FLD) represents an important cause for degenerative disruption and is increasingly recognised as an important cause (up to 25%) of degenerative dementia among late-middle-aged individuals (American Psychiatric Association, 1994; Cummings, 2000; Miller, 2001). Its true prevalence is unknown because the disorder has often been misdiagnosed, as indicated by the study reported by Mendez et al. (1993), who found that 18 of 21 patients with pathologically confirmed Pick's disease had been erroneously diagnosed as Alzheimer Disease while they were alive. Moreover, many patients with FLD are never seen in dementia centres because of the predominantly psychiatric presentation of the illness (Miller, 2001). The widespread definition of FLD concern the behavioural alterations, which are the core diagnostic features in all the guidelines: early loss of personal and social awareness, that are associated with disinhibition, mental rigidity and inflexibility, distractibility, impulsivity. Also, emotional unconcern (lack of empathy and sympathy, emotional indifference and remoteness) and impersistence are very frequent; all of them distinguish nearly 80% of FLD (Duara, et al., 1993; The Lund and Manchester Groups, 1994; Levy et al., 1996; Miller et al., 1997; Sjogren, 2000: *Clinical and physiopathological aspects of FTD.* Oral Communication- International Congress: Alzheimer's disease versus other degenerative dementias: from biology to treatment. Perugia, Italy).

Although relatively easy to define, vascular dementia has proved difficult to diagnose (Ransmayr, 1998) because of the wide variability in mechanisms, brain changes, syndromes, and courses of the disease. At present, the current criteria for vascular dementia such as that in the DSM-IV (Mendez, et al., 1993) identify a heterogeneous group and may have biased findings in research studies and clinical trials. In 1993, the NINDS-AIREN Intemational Work Group (Roman et al., 1993) proposed diagnostic criteria for vascular dementia focusing on the delineation of the clinical subtypes of vascular dementia, consistent with the heterogeneity of dementia syndrome due to cerebral vascular disease. Accordingly, the main subcategories of ischaemic vascular dementia include multi-infarct dementia, strategic single infarct dementia and small-vessel disease with dementia. Subcortical vascular dementia, which relates to small vessel disease and hypoperfusion resulting in focal and diffuse ischemic white matter lesions and incomplete ischemic injury (Corbett et al., 1994), particularly applies to the prefrontal subcortical circuit, including the prefrontal cortex (Corbett et al., 1994).

Deterioration of the frontal lobe is reflected by the dysexecutive syndrome that seems to be the core feature of subcortical vascular dementia (McPherson and Cummings, 1996; Looi and Sachdev, 1999). Memory impairment and attentional deficits are also apparent, and patients often experience mood changes such as depression, personality changes and emotional lablilty. In particular, these behavioral symptoms can be a major cause of stress, anxiety and concern for caregivers, and frequently lead to the institutionalization of patients.

The purpose of a singular study (Moretti et al., 2004) was to analyze the performance of subcortical vascular patients and subjects affected by frontal lobe dementia, in a set of attentional, executive, and behavioural tasks: the FLD and the subcortical vascular dementia involve the frontal neural pathways. We expect to find an earlier decay in the cortical FLD group than in vascular dementia, either in behavior either in cognitive performances.

Participants were eighty outpatients of the Cognitive Disturbances Section of the Neurological Clinic of the University of Trieste, aged 60-75 years, satisfying the fourth edition of the Diagnostic and Statistical Manual of Mental Disorders (DSM-IV) for dementia (APA, 1994); forty patients satisfied the criteria for probable FLD in accordance with the Lund and Manchester Criteria (The Lund and Manchester Groups, 1994), and forty patients satisfied the criteria for probable vascular dementia, in accordance with the National Institute of Neurological Disorders and Stroke and the Association Internationale pour la Recherche et l'Enseignement en Neurosciences (NINDS-AIREN) criteria (Roman, *et al.*, 1993). Patients with a history of significant head trauma, alcoholism, movement disorder, or any other condition known to impair frontal lobe function were excluded. Patients with significant impairments on tests of semantic memory or altered linguistic pattern were also excluded.

All patients underwent a standardized baseline assessment that included a detailed history, a physical examination, laboratory test, and psychiatric evaluations. All the patients had undergone a CT of the head and an MRI evaluation. The neuroimages of FLD patients showed clear signs of frontal atrophy, without asymmetric preferences and with relatively little change of the other cortical and subcortical regions. On the other hand, a patient was diagnosed as having subcortical vascular dementia when the CT scan showed moderate to severe ischaemic white matter changes and at least one lacunar infarct (Erkinjuntti et al, 1987). Patients were not included in the study if they showed signs of non-lacunar territorial infarcts, normal pressure hydrocephalus, cortical hemispheric large vessel strokes or lobar haemorrhages.

Patients were divided into two groups of forty, matched for education level, one is composed by patients with frontal lobe dementia, and the other of patients with subcortical vascular dementia.

All patients were followed for 12 months, with periodical neurological and neuropsychological examinations. Visits were scheduled to take place one, three, six, and 12 months after. A complete neuropsychological examination was conducted at baseline, and at 12 months.

The trial was conducted in accordance with the Helsinki Declaration and in accordance with the Ethics guidelines of the Institute. Treatment compliance was assessed by caregivers, who were instructed to report any problems.

Global cognitive function was assessed using Mini-Mental State Examination (MMSE) (Folstein et al., 1975) at each visit. In addition, executive and planning function were assessed using the Ten Point Clock Test (Manos, 1997), the Proverbs Test (Gorham, 1956), and the Stroop Test (Stroop, 1935) at each visit.

Behavioral performance was assessed using different tests, reliable and validated, widely known by the Literature: the NeuroPsychiatric Inventory (NPI) (Cummings et al., 1994) (minimum score=0 which indicates no behavioral disturbances; maximum score=144), the Clinical Insight Rating Scale (CIR) (Ott et al., 1996) (lowest score=0, which indicates a perfect insight; highest score=8, no insight), and the Cornell Scale for Depression in Dementia (Alexopoulos et al., 1988) (lowest score=0, no depression, highest score=38) at every visit. The Behavioral pathology in Alzheimer's Disease Rating Scale (BEHAVE-AD) (Reisberg et al., 1987) (lowest score=0, no behavioral alteration; highest score=75) was given at baseline and at 14 months. The last scale, originally thought for Alzheimer's Disease, could be a successful tool, reliable and valid, in the assessment of behavioral alteration of different etiologies (Mendez et al., ). The factors measured by this scale includes agitation/anxiety of upcoming events or other events, psychosis (delusions of theft, suspiciousness, hallucinations –visual, verbal or tactile), aggression (verbal aggressiveness, physical treats, fears), depression and activity disturbances (wandering). In addition, new prescriptions of neuroleptics and or benzodiazepines were recorded and compared between the two treatment groups (indicated by a binomial scale where 1 = intake and 0 = no intake). Caregiver's stress was evaluated at each visit by the Relative Stress Scale (Greene et al., 1982).

Statistical analyses were performed using the Statistical Package for Social Sciences (SPSS, version 10.0). Changes from baseline to 20 months between Group A and B were compared using the 2-sample Wilcoxon Test and the Marginal homogeneity test. Results are presented as means with standard deviations and p-values are presented where appropriate (Moretti et al., 2004).

Forty patients with a diagnosis of probable FLD and forty patients with a diagnosis of subcortical vascular dementia were included in this study (Table 1). Their mean age was 69. 8 $\pm$ 2.5 years, and they have a mean educational level of 8.8 ± 2.2 years. The FLD patients' first symptoms have began for 6 months (mean duration 7. 32 $\pm$ 1.56months), and were described by caregivers as social misconduct, irritability, apathy, abulia, dis-inhibition, lack of concentration. The vascular dementia patients' first symptoms could be dated to about 10 months before the visit (mean duration 10 ± 2.3). A two-tailed t-test has been done for each efficacious variable, and results have been reported in Table 1.

**Table 1. Baseline characteristics of patients in the two treatment groups**

| | Group A (FLD) | | t-test |
|---|---|---|---|
| No of patients | 40 | | Ns |
| Gender (male/female) | 29/11 | | Ns |
| Age, years (mean ± SD) | 63.13 ± 3.65 | | P<0.05 |
| Education level, years (mean ± SD) | 8.57 ± 6.23 | | Ns |
| Handedness | Right | | Ns |

**Table 2. Baseline characteristics of patients in cognitive and behavioral scores; results are presented as mean and SD**

| Outcome measure | Group A FLD | Group B VaD | P |
|---|---|---|---|
| MMSE | 24 ± 1.56 | 18.07 ± 1.53 | p<0.01 |
| TPC | 2.53 ± 1.41 | 3.2 ± 0.86 | P<0.01 |
| Proverb Test | 8.27/20 ±1.75 | 11.53/20 ± 1.19 | P<0.05 |
| Stroop Test | 81 seconds± 5.39 | 68.4 seconds ± 1.96 | P<0.01 |
| | 6.47 mistakes± 1.6 | 4.8 mistakes ± 0.86 | P<0.01 |
| NPI | 33 ± 4.69 | 22.87 ± 2.75 | P<0.01 |
| CIR | 6.27 ±0.96 | 1.8 ± 0.68 | P<0.01 |
| Cornell Depression Scale | 12.13± 1.77 | 27.6 ± 4.4 | P<0.01 |
| BEHAVE-AD | 26.27 ± 4.7 | 23.07 ± 3.69 | P<0.01 |
| RSS | 46.6 ± 10.74 | 31.33 ± 4.35 | P<0.01 |

Brain CT-scans or MRI scans were available for all patients. MRI images for 40 VaD patients were confirmed as satisfying criteria for small vessel disease represented by at least two white matter lacunae, or periventricular white matter hyperintensities in the left or right frontal parietal regions. The MRI evaluations for 40 FLD patients showed clear signs of frontal atrophy, without asymmetric preferences and with a relative spare of the other cortical and subcortical regions.

At baseline, the two groups differ significantly in cognition and behavioral pattern: a Marginal Homogeneity test shows a statistical difference in cognitive ability, as far as MMSE is concerned (asymptotic significance-2-tailed, $p<0.001$), in executive function, as measured by Ten Point Clock test (asymptotic significance-2-tailed, $p<0.05$), in abstract reasoning, as measured by Proverb test (asymptotic significance 2-tailed, $p<0.001$), and in attention, as measured by Stroop test, time and number of mistakes (asymptotic significance 2-tailed, $p<0.001$, respectively), and in behavioural performances, as measured by NPI and BEHAVE-AD scores (asymptotic significance 2-tailed, $p<0.001$, respectively), depression score, as showed by Cornell scale (asymptotic significance 2-tailed, $p<0.001$), and insight in their clinical situation, as measured by CIR (asymptotic significance 2-tailed, $p<0.001$) (Table2).

Both Groups had significant changes in cognitive and behavior characteristics during follow-up; the caregivers referred that group A (FLD) patients exhibited personality changes, behavioral withdrawal, lack of insight and disinhibition. Moreover, the patients showed apathy and indifference towards the world, with total loss of "social rank", sometimes resulting in overt aggressiveness. On the contrary, the caregivers of patients in group B (VaD) referred a diminishment of executive function and lack of set shifting, and patients being more rigid and apathic.

At 12 months (Table 3), results obtained by the frontal patients could be summarised as follows: they showed statistically significant deteriorations in cognition performances, as stated by a decrease in MMSE ($p<0.01$, according to a Wilcoxon Signed Ranks Test), in executive functions, as evidenced by the scores obtained in TPC ($p<0.05$), in abstract reasoning, as showed by Proverbs scores ($p<0.01$) and in attention, as showed by Stroop number of mistakes ($p<0.01$). On the other hand, there is a general worsening of behavioural performances as stated by the NPI and BEHAVE-AD scores ($p<0.01$ respectively), and by a slight reduction of depression scale ($p<0.05$). There is a total lack of the insight in their clinical conditions, as showed by a reduction of CIR scale, ($p<0.05$). The caregiver' burden is heavier, as stated by the increase of the RSS ($p<0.01$).

Thus (Table 3), patients in Group B showed statistically significant deteriorations in cognition performances, as stated by a decrease in MMSE ($p<0.01$), in executive functions, as shown by the scores obtained in TPC ($p<0.01$), in abstract reasoning, as shown by Proverbs scores ($p<0.01$) and in attention, as indicated the number of mistakes on the Stroop task ($p<0.05$). There was an increase of the NPI and BEHAVE-AD scores ($p<0.01$ and $p<0.05$, respectively), and an increase of depression scores ($p<0.01$). There was a general stability of the insight in their clinical conditions, as showed by a reduction of CIR scale. The caregiver's stress was increased, as stated by RSS' scores ($p<0.05$).Table 3 shows the final differences in cognitive and behavioural rating scales in the two groups of patients. At 12-months, the two groups differed significantly in scores for cognition and behavioural pattern: a Marginal Homogeneity test showed a statistical difference in cognitive ability, on the MMSE ($p<0.001$), but not in executive function (Ten Point Clock test), in abstract reasoning (Proverb test, $p<0.001$), and in attention (Stroop test, time and number of mistakes, $p<0.05$, and $p<0.001$, respectively). As far as behavioural performances are concerned, there were significant differences in NPI and BEHAVE-AD scores ($p<0.01$, and $p<0.001$, respectively), in scores for depression (Cornell' scale, $p<0.001$), and in insight in their clinical situation (CIR, $p<0.001$) (Moretti et al., 2004).

**Table 3. Changes in outcome scores at 12 months compared with baseline in the two groups (follow-up scores minus baseline scores)(mean ± SD) and compared within the two groups (mean ± SD)**

| Outcome measure | Group A FLD(baseline) | Group B VaD (baseline) | Group A FLD(within group) | Group B VaD (within groups) |
|---|---|---|---|---|
| | | mean sd | mean sd | |
| MMSE | -6.07¶¶ ± 0.93 | - 8.8¶¶±0.1 | 17.93 ±1.83 | 9.27** ±1.44 |
| TPC | - 2.42¶ ± 0.1 | -2.21 ¶¶ ±0.1 | 0.11±1.42 | 0.99 ns±0.83 |
| Proverb Test | -3.2¶ ± 0.65 | -2.73¶¶ ±0.13 | 1.6 ±1.4 | 8.8** ±1.06 |
| Stroop Test | -7.67(sec) ¶ ±0.45 | 3.24(sec) ¶±2.87 | 73.33(sec) ±5.85 | 71.24 (sec) **±5.47 |
| | 6.53(mistakes) ¶¶±1.2 | 1(mist.) ¶±0.12 | 13.41(mist.)±2.38 | 5.8(mist.) **±0.94 |
| NPI | 7.64¶¶ ±1.2 | 4.3 ¶¶ ±0.3 | 40.64 ±3.78 | 27.17** ± 2.22 |
| CIR | 1.41¶ ±0.21 | 0.33 ns ±0.14 | 7.81±1.01 | 3.13** ± 0.74 |
| Cornell Depression Scale | - 3.56¶ ±0.67 | 7.57¶¶±1.42 | 8.57 ± 2.61 | 35.17** ± 2.47 |
| BEHAVE-AD | 8.53 ¶¶±1.67 | 3.41¶ ±2.51 | 34.8 ± 2.24 | 26.57** ± 2.21 |
| RSS | 21.34¶¶ ±2.56 | 5.06¶ ±1.27 | 67.94 ± 7.73 | 36.39* ± 3.01 |

$p < 0.05$ and ¶¶$p < 0.01$ change compared with baseline.
$p < 0.05$ and **$p < 0.01$ change compared within groups.

To our knowledge, this is the first study that compares cognitive and behavioral aspects of frontal lobe degeneration and subcortical vascular deterioration (Moretti et al., 2004). Even considering the limited simple size, it must be noted that both the groups have been selected at very initial stages of pathology (6 months was the FLD mean duration of symptoms before the first visit, 10 months was the vascular dementia mean duration of symptoms before the first visit). The vascular dementia group is eight years older than the frontal dementia group. The number of participants is quite small and, obviously, a group matching would reduce the number of participants even further. However, it is known that frontal dementia clinical syndrome usually merges earlier than vascular dementia; the most common age of onset of FLD is the sixth decade (Munoz and Kertesz, 2003). Psychometric testing has been found a useful diagnostic tool to diagnose FLD patients, but the Mini Mental Status Examination may be normal in many initial examinations (Kertesz and Munoz, 1998). The normal scores obtained in frontal tests measuring spatial working memory and planning tasks, in the early stages of the FLD syndrome suggest the dorsomedial frontal or orbitofrontal cortex as the initial site of pathology in these patients (Rahman et al., 1999).

What claims to our eyes is that, even if our FLD group is 2-3 SDs younger than vascular group, it deteriorates faster than the vascular group. This observation limits the possibility that all the differences we observed do not simply reflect overall dementia severity or average age of onset (Moretti et al., 2004).

The purpose of our study was to analyze the performance of subcortical vascular patients and subjects affected by frontal lobe dementia, in a set of attentional, executive, and behavioural tasks: the FLD and the subcortical vascular dementia involve the frontal neural pathways (Moretti et al., 2004). We hypothesize, based on previous research, that an earlier decay occurs in the cortical FLD group than in vascular dementia, either in behavior either in cognitive performances.

Among all the five, or six circuits defined as cortical-basal ganglia circuits (Middleton, 2003), three frontal-subcortical circuits have been involved in cognition and behaviour (Miller, 2000).

A common feature of all these circuits is their unification of different parts of the frontal cortex with the basal ganglia and the thalamus in closed circuits with open elements that receive input from and project to other regions outside the loops. The hypothesized function of the orbitobasal circuit is the correction of social control as the deficits seen with the selective disruption are disinhibition, impulsiveness, confabulation and antisocial conduct. The dorsolateral net supports attention and alternative programming capacities, as well as working memory (area 46) programs. Its deficit causes poorly focus attention, marked alteration of working memory, poor organization, defective planning capacities, lack of insight, and abnormal expression of empathy and sympathy. The medial frontal circuit might be involved in energy, motivation and selectivity of thoughts, while its disruption may cause apathy, abulia and marked depression. Within each of the circuits there are two pathways: a direct, which links the striatum with the globus pallidus interna/substantia nigra complex and an indirect pathway, which projects from striatum to globus pallidus externa, to subthalamus, to globus pallidus interna/substantia nigra to thalamus. A direct involvement of the cortical (decisional) layers in frontal dementia (Moretti and Torre, 2003), in contrast with the prominent and widespread involvement of the subcortical pathways (refinement and corrections programs), may be the reason for the different cognitive and behavioural profiles of the two groups: the frontal patients we described have more difficulties in abstract reasoning, in focusing attention, and in implementing strategies to solve problems. In contrast, they exhibit more profound behavioural alterations, in personality and social conduct, loss of basic emotions and social embarrassment, selfishness, and disinhibition. The patients showed moderate depression and a total lack of insight in their clinical condition. On the contrary, our patients with subcortical vascular dementia have poor general cognitive functions, a high level of insight, an important depression and apathy as the principal and most salient characteristic of their behavioral conduct. Dys-executive syndrome (which better characterize subcortical vascular dementia; Cannatà et al., 2002) was not so evident when compared to results obtained by our patients with frontal lobe dementia. Moreover, subcortical vascular dementia encompasses both cognitive impairment, and also a wide range of behaviour disturbances (Cannatà et al., 2002; Traykov et al., 2002). A well-conducted study (Aharon-Peretz et al., 2000) indicated that patients with mild subcortical VaD are more disturbed on behavioural aspects that are influenced by frontal and subcortical mechanisms than AD patients. This suggests that in addition to neuropsychological assessment, quantification of the personality behaviour disorder is important for standardizing the diagnosis of subcortical vascular dementia and distinguishing it from any other dementias.

The results obtained from the study by Moretti et al. (2004) are limited due to the small number patients enrolled. In future research questions need to be considered. For example, is frontal lobe dementia a unique entity, or should it be considered as three different variants (an orbito-basal, a dorso-lateral, which is the most frequent aspect in the selected patients, and a medial-frontal) with different anatomical, neurochemical pathways, therefore requiring different therapies?

In: Basal Ganglia and Thalamus
Editors: R. Moretti, P. Torre and R. M. Antonello
ISBN 978-1-60692-198-2


*Chapter 15*

# The Thalamus: Its Role in Cognition and Behaviour

***Rita Moretti, Paola Torre, Rodolfo M. Antonello, Antonio Bava****
* Dip. Fisiologia e Patologia, Università di Trieste, Italy

The BG complex consists of several component structures; the Striatum, the Sub-Thalamic Nucleus (STN), the Globus Pallidus internal (GPi) and Globus Pallidus external (GPe) segments, and the Substantia Nigra (par compacta-SNPc and par reticula- SNpr).To involve the thalamus in a direct effect on cognition and behaviour, one should take into consideration its integrative role among basal ganglia, as seen below (Figure 16).

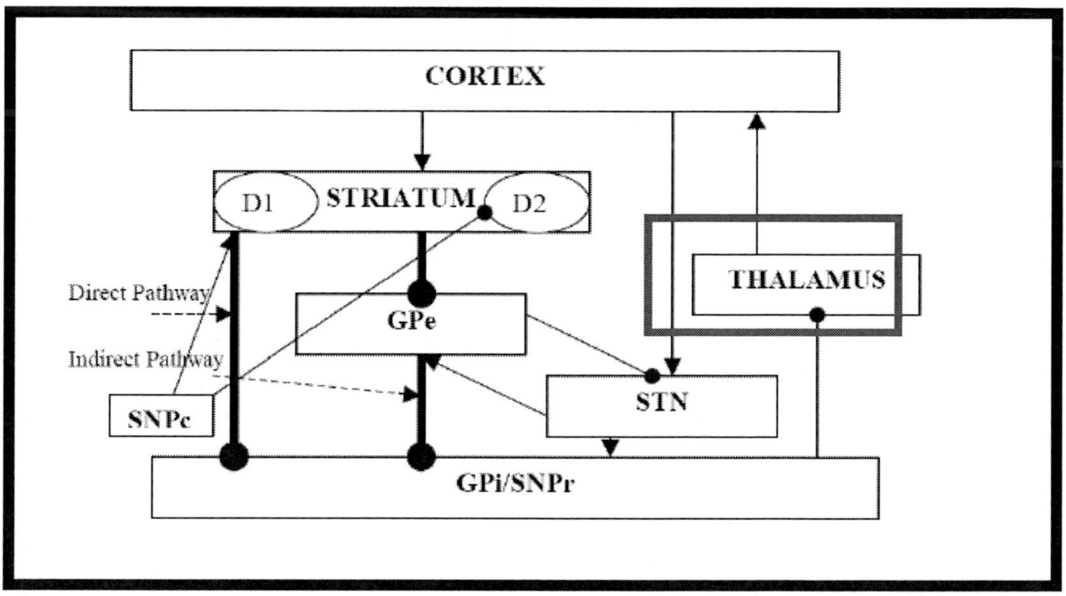


Figure 24. Basal ganglia and thalamus net.

A significant neuroantomic fact about the primate basal ganglia complex is their involvement in five parallel cortical –thalamic loops namely two motor loops (Skeletomotor and Occulomotor), and three non-motor loops.

The non-motor loops include a Dorso-lateral prefrontal-cortex loop, a Lateral Orbito-frontal loop and an Anterior Cingulate loop.

The Dorso-lateral Pre-frontal loop (DLPFC (Area 9 an 10) – basal ganglia (Caudate Nucleus - Gpi –SNPr) –Thalamus (VA, MD) –DLPFC) is believed to be involved in Executive functions, planning and working memory.

The Lateral Orbitofrontal Loop (LPFC – basal ganglia (VM Caudate –GPi- SnPr) – Thalamus – LPFC) mediates empathetic and socially appropriate responses. Damage to this loop produces irritability, lack of empathy and is implicated in Obsessive Compulsive Disorder (OCD).

The Anterior Cingulate Loop (Ant.Cing.Gyrus –BG (Ventral Striatum -RostroMedial Palladium – SNPr) – Thalamus (MD thalamic nucleus) – ACC) is believed to be involved in communicating reinforcement signals from ventral tegmental area (VT) and SNpc.

Damage to this loop may result in akinetic mutism (profound defects in movement initiation).

The concept of "parallel" processing within basal ganglia pathways relies, in part, on the fact that distinct thalamic relay nuclei, that receive basal ganglia output, project to specific motor, premotor, and prefrontal cortical areas (Haber, 2003). Thalamic relay nuclei, once thought to simply passively transfer information from afferent systems to cortex, are now known to play an important role in changing the dynamics of cortical processing by setting up different oscillation patterns of frequency and synchrony (Sherman, 1996; Haber, 2003). For example, the resting tremor in PD is thought to be the consequence of changes in thalamic firing frequency due to an increased inhibition of the pallidal input to the VA/VL cell groups. This inhibition leads to a hyperpolarization of thalamic cells and changes in its oscillatory firing pattern to cortex (Pare et al., 1990; Haber, 2003).

Even more fascinating is the large corticothalamic projection. Indeed, this projection is 10 times as great as the thalamocortical projection. This projection controls the oscillatory firing pattern of the thalamus (Jones, 1998; Contreras and Steriade, 1998; Steriade, 1999; Haber, 2003), through a feedback mechanism that inhibits relay neurons (Destexhe et al., 1999; Bal et al., 2000; Haber, 2003). Thus, the relationship between the thalamocortical and corticothalamic projection systems of the ventral and MD thalamic nuclei is likely to have an important impact on how information is processed for each basal ganglia circuit. Furthermore, this relationship is likely to have important implications on how information transfer can take place between circuits (Haber, 2003).

There are two components to the corticothalamic projection:

1) a reciprocal component, in which the cortical area receiving the thalamic input projects back to the same thalamic area;
2) a nonreciprocal component, in which a given cortical area projects to a region of thalamus that does not receive its input.

Thus, corticothalamic terminal fields to a given thalamic region are derived from a wider range of cortical areas than the reciprocal thalamocortical output (Sherman, 1996; Haber, 2003*)*. The nonreciprocal component of the projection arises from large cells in layer V, while the reciprocal component is derived primarily from small cells in layer VI (see data and Literature in Haber, 2003).

The nonreciprocal neurons of layer V are large rapidly conducting cells, which may therefore be responsible for changing the thalamo-cortical oscillatory patterns.

This change in firing pattern would, therefore, effect a different part of cortex providing a mechanism for cortex to synchronize thalamic oscillations in distal cortical regions (Destexhe et al., 1999). In this way, specific thalamic relay nuclei modify and facilitate information flow across functional regions of cortex via specific thalamic nuclei, and these thalamic nuclei can modify that information (Haber, 2003). Thus, the interactions between cortical and thalamic activity may act in a general way to influence global cortical activity. AlternativeIy, it may function in a specific feedforward mechanism that facilitates information transfer from one cortical area to another. This has been a proposed mechanism by which primary sensory cortex can influence "higher" cortical association areas *(*Jones, 1998; Contreras and Steriade, 1998; Steriade, 1999; Haber, 2003*)*.

Corticothalamic projections to specific VA/VL and MD sites are more widespread than the VA/VL and MD thalamocortical projections. Based on separate experiments that demonstrate the thalamocortical projections and the corticothalamic projections of the basai ganglia thalamic relay nuclei, it can be hypothesized that these nuclei also receive both reciprocal and nonreciprocal corticothalamic inputs (McFarland and Haber, 2002; Haber, 2003). For example, VL has reciprocal connections primariIy with caudal motor areas.

In addition, it receives a nonreciprocal cortical projection from more rostral molar areas. VA is reciprocally connected with rostral motor areas, and dorsolateral prefrontal areas, but it also receives a non-reciprocal corticothalamic projections from medial areas, as well as lateral orbitofrontal areas. Reciprocal connections with MD are similarly limited, involving primarily dorsolateral prefrontal and lateral orbitofrontal areas. However, nonreciprocal MD inputs also arise from medial, limbic-related, prefrontal areas, such as 24a/b and 32. Thus, the *cor*tico-thalamo-cortical circuit relays information from different basal ganglia loops forming a feed-forward circuit associated with limbic, cognitive, motor planning, to motor execution.

Information from each thalamic relay nuclei is conveyed back to the cortical region of origin, providing the last link in the parallel and segregated circuits. However, since there is also a nonreciprocal corticothalamic input to each of these thalamic nuclei, the incoming basal ganglia information from each circuit is mixed with cortical inputs from a different circuit (Haber, 2003). Since the nonreciprocal input to each thalamic nucleus is derived, at least in *part, from* more rostral and medial cortical areas, the flow of information from *cortex* to thalamus is from limbic and cognitive areas into premotor and motor regions of thalamus. Therefore, although the last link to *cortex* is through the basal ganglia and follows a parallel pathway, the bidirectional connections between cortex and thalamus may bave a great infiuence on how segregated the infomation is frolli each loop. The information that the thalamic relay nuclei convey to cortex is, therefore, not only effected by the parallel pathway through the basal ganglia *struc*tures, but is also modified by the non-reciprocal corticothalamic pathway, resulting in an integrated feedforward processing similar to that seen in sensory systems (McFarland and Haber, 2002; Haber, 2003).

In: Basal Ganglia and Thalamus
Editors: R. Moretti, P. Torre and R. M. Antonello
ISBN 978-1-60692-198-2


Chapter 16

# The Thalamus and the Clinical Reality

***Rita Moretti, Paola Torre and Rodolfo M. Antonello***

According to pathological data, the thalamus is usually divided into four territories:

1) the *anterior*, *tuberothalamic* or *polar* territory (involvement of the anterior group of thalamic nuclei, the anterior part of the intralaminar nuclei and the mamillo-thalamic tract (MTT); except for its most anterior part, dorsomedian nucleus is usually spared). This territory, supplied by the tuberothalamic artery (polar artery of Percheron, anterior optic arteries of Duret, premamillary pedicle of Foix and Hillemand), accounts for about 12 to 18 per cent of all thalamic infarcts (Bogousslavsky et al., 1988-A; Saez de Ocariz et al., 1996). The most frequent etiology is cardio-embolism (Carrera et al., 2004).
2) the *paramedian* territory (involvement of the median part of the thalamus including the intralaminar (IL) nuclei and most of the dorsomedian (DM) nucleus). This territory, supplied by the paramedian arteries (retromamillary or thalamoperforate pedicle of Foix and Hillemand), accounts for about 22 to 35 per cent of all thalamic infarcts (Bogousslavsky et al., 1988-A; Saez de Ocariz et al., 1996). The most frequent etiology is cardio-embolism (Carrera et al., 2004).
3) the *inferolateral* or *thalamogeniculate* territory (involvement of the ventromedial (VM), ventrolateral (VL) and ventro-posterolateral (VPL) nuclei). This territory, supplied by the thalamogeniculate arteries (thalamogeniculate pedicle of Foix and Hilleman , or infero-external optic artery of Duret), accounts for about 45 to 50 per cent of all thalamic infarcts (Bogousslavsky et al., 1988-A; Saez de Ocariz et al., 1996). The most frequent etiology is microangiopathy (Carrera et al., 2004).
4) the *posterior choroidal* territory (involvement of most of the pulvinar nucleus). This territory, supplied by the medial and lateral branches of the posterior choroidal arteries (Cosson et al., 2003) (posteromedial and posterolateral system of Castaigne et al., 1981), accounts for about 7 to 9 per cent of all thalamic infarcts. The most frequent etiology is microangiopathy (Carrera et al., 2004).

The anterior artery usually originates from the posterior communicating artery (PCoA), whereas paramedian arteries originate from the first part of the posterior cerebral arteries (P1) and thalamogeniculate and posterior choroidal arteries originate from the second part of the posterior cerebral artery (P2). Usually, the paramedian arteries originate directly from P1 on each side but, in one third of brains, they originate from a single pedicle (Castaigne et al., 1981; Percheron, 1976). In one third of hemispheres, the tuberothalamic artery lacks and the paramedian arteries supply the anterior territory (Castaigne et al., 1981; Percheron, 1976).

Recently, three variant territories have been described overlapping the four "classical" territories (Carrera et al., 2004).

Infarcts in these territories probably result from borderzone infarcts or variation of "classical" thalamic arteries. Infarcts restricted to the thalamus account for 11 per cent of all vertebro-basilar infarcts. The thalamus is involved in 27 per cent of vertebro-basilar infarcts (Bogousslavsky et al., 1988-B) and in about one third of patients with basilar artery occlusion (Labauge et al., 1981). One third of patients with occipital infarct show concomitant thalamic infarction (Yamamoto et al., 1999).

It should be taken into account that clinical symptoms are tightly due to the anatomical involvement of different thalamic areas. Therefore, Bogousslavsky and Carrera (2005) divided the casistic of Lausanne into four patterns of clincial presentation: the anterior, the paramedian, the inferolateral and the posterior syndromes.

In human, the anterior nuclei group is divided into three parts: the largest: the ventral nucleus, the medial nucleus and the smaller dorsal nucleus (Bogousslavsky and Carrera, 2005). The ventral and medial anterior nuclei are sometimes grouped together as the principal anterior nucleus. Anterior nuclei are the relay of mamillothalamic tract (or Vicq d'Azyr tract) in thalamus. Anterior nuclei are reciprocally connected to the anterior limbic system including cingulate gyrus, hippocampus, parahippocampal formation, including entorhinal cortex, retrosplenium and orbitofrontal cortices (Yakovlev and Locke, 1961). Other connections to the medial prefrontal cortices, posterior area of the neocortex and ventral pallidum have been described (Schmahmann and Pandya, 1990).

Patients with infarct in the anterior territory frequently show a perseverative pattern in thinking and speech with inappropriate maintenance of category but also in all memory and executive tasks with increased sensitivity to interferences (Bogousslavsky and Carrera, 2005).

In addition, many patients show a superimposition and "telescoping" of unrelated information with parallel expression of mental activities that we have called palipsychism from Greek palin (again) and psychê (soul) (Ghika-Schmid and Bogousslavsky, 2000; Bogousslavsky and Carrera, 2005).

One of the patients described (Ghika-Schmid and Bogousslavsky, 2000; Bogousslavsky and Carrera, 2005) correctly solved a calculation problem while simultaneous speaking of his garden and describing his job or correctly explaining the sequences of a journey together with giving details on familial problems and financial issue. The output speech after anterior territory infarct is characterized by grammatically correct phrases but unpredictable topic shifts which are, in most cases, not dependant on thecurrent activities but usually intrusions of previous topics introduced by the patient or the examiner (Chatterjee et al., 1997; Bogousslavsky and Carrera, 2005).

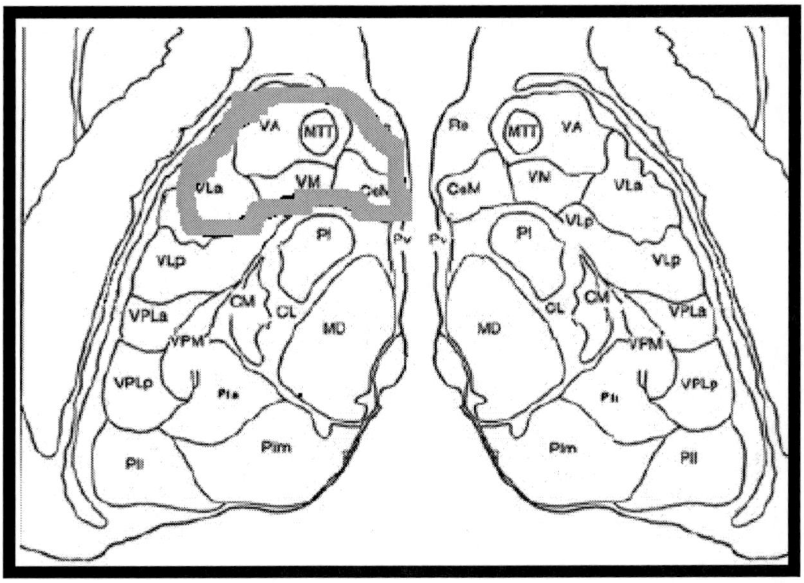


Figure 25. A cartoon for an ideal representation of the anterior thalamic nuclei (modified by the Authors).

Some patients show a spectacular pattern with "impersistance de la pensée" which may be reminiscent of the "psychotic" speech found in schizophrenic patients. These speech disturbances are mainly found in unconstrained speech, for example in comment of news or historical facts and in definition of words, which contrasts with a relative maintenance in performing automatic series or reciting poems(Chatterjee et al., 1997; Bogousslavsky and Carrera, 2005). The output speech difficulties reappear when the patient is asked to explain the poem that was recited before and are consistent with a confusion of ideas rather than aphasia per se (Chatterjee et al., 1997; Bogousslavsky and Carrera, 2005).

The role of thalamo-cortical projections, especially to the frontal lobes is highlighted by the similar behavior changes found in a patient with an isolated infarct of the genu of the left internal capsule adjacent to the anterior thalamus (Tatemichi et al., 1992; Bogousslavsky and Carrera, 2005). The speech output was described with grammatically correct phrases which were however often strung together in a nonsensical or illogical fashion and occasionally punctuated by paraphasias or neologism. Perseverations were also found. SPECT images showed a diffuse hypoperfusion of the left hemisphere, especially prominent in the medial and lateral frontal lobes.

Bogousslavsky and Carrera (2005) infer that these behaviour changes found after anterior thalamic infarct in the territory of the anterior arteries with perserverations and frequent topic shifts in output speech results from a disconnection of the ipsilateral cortex, mainly the medial and lateral frontal regions, which seems to be coincident to what we have found and previously described (Moretti et al., 2003 C).

Aphasia was never found to be the main neuropsychological dysfunction after anterior nuclei infarction and, when present, is usually consistent with a moderate form of transcortical motor aphasia associated with hypophonia and dysarthria (Ghika Scmid and Bogousslavsky, 2000; Bogousslavsky and Carrera, 2005) . Aphasia usually occurs with left or bilateral infarcts, although Bogousslavsky et al. (1986) found four patients (two left-handed and two right-handed) with infarct limited to the right anterior territory with transcortical motor aphasia (Ghika Scmid and Bogousslavsky, 2000; Bogousslavsky and Carrera, 2005). Spontaneous speech is usually reduced with lack of speech initiation and short sentences (usually less than 10 words). Occasional phonemic paraphasias are found ("anassin" instead of "assassin") as well as semantic paraphasias ("clock" instead of "watch") or neologisms (Bogousslavsky and Carrera, 2005). Repetition and comprehension are only rarely impaired. Phonemic paralexias as well as calculation and writing impairment are rarely found (Bogousslavsky et al., 1986; Ghika Scmid and Bogousslavsky, 2000; Bogousslavsky and Carrera, 2005).

Anterograde memory deficit is a constant finding after anterior infarcts resulting from interruption not only of amygdalofugal tract but mainly of mamillo-thalamic tract (MTT) and its projections to the cingulate gyrus and other structures of the limbic regions such as hippocampus, parahippocampal formation, entorhinal cortex, orbitofrontal and medial prefrontal cortices and subcortical structures including the mamillary bodies and

amygdala (Bogousslavsky and Carrera, 2005).

Amnesia, which can occur without other behavior or neuropsychological disturbances if the lesion is restricted to the MTT, is usually more severe after anterior infarcts restricted to the MTT than after more posterior lesions involving the dorsomedian or intralaminar nuclei. Clinically, anterograde amnesia with better performance for recognition is a constant finding (Bogousslavsky et al., 1986). On the contrary, short-term memory is usually spared (Bogousslavsky and Carrera, 2005). Severe impairment in explicit recall of new facts and events, with intact word-complexion priming and remote memory were described after bilateral anterior infarct (Malamut et al., 1992; Bogousslavsky and Carrera, 2005). Memory dysfunction may be linked to a selective field of memory: in some cases, visuospatial memory impairment was reported to be prominent after right lesion31,32,33 and verbal memory impairment to be prominent after left lesion (Bogousslavsky et al., 1986; Clarke et al., 1994; Bogousslavsky and Carrera, 2005). Numerous intrusions, perseverations and false recognition are frequently found during neuropsychological tests. Intrusions during the recall of a word list (provoked confabulations) are more frequent than during spontaneous speech (spontaneous confabulations) (Ghika Scmid and Bogousslavsky, 2000; Bogousslavsky and Carrera, 2005). In some patients, different authors observed fantastic and bizarre confabulations which looked like those found in patients with left medial frontal lobe infarct in the territory of the anterior cerebral artery and suggest a disconnection of the orbital and medial frontal cortex (Kapur and Coughlan, 1980; Benson et al., 1996; Schnider et al., 1996; Bogousslavsky and Carrera, 2005).

Paramedian nuclei consist mainly of dorsmedian nucleus and intralaminar nuclei. The dorsomedian nucleus is divided into two parts*: the anteromedial magnocellular nucleus*, which receives projections from the amygdala, is reciprocally connected with the anterior and medial prefrontal cortices, with the mediobasal nucleus, the ventral pallidum and the enthorhinal cortex and *the posterolateral parvocellular mediodorsal nucleus* has also reciprocal connections to the dorsolateral and dorsomedial prefrontal cortices (Bogousslavsky and Carrera, 2005).

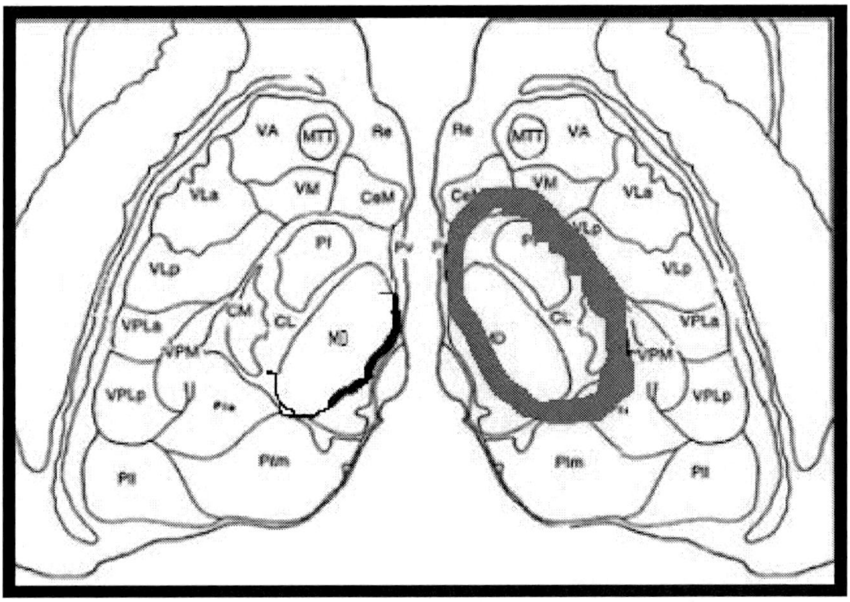


Figure 26. A cartoon for an ideal representation of the paramedian thalamic nuclei (modified by the Authors).

The intralaminar nuclei consist of the parafascicular, the centromedian, the central medial, the paracentral and the central lateral nuclei. Smaller nuclei of the "midline" such as the paraventricular, rhomboid and reunions nuclei are also included in the intralaminar nuclei. Intralaminar nuclei are divided in two parts: *the caudal part*, which includes the centomedian and parafascicular nuclei, has reciprocal projections to the motor, premotor and supplementary motor cortices, to the substantia nigra pars reticulata and internal globus pallidus, and *the rostral part* comprises the central medial, the paracentral and the central lateral nuclei. The central medial nucleus projects mainly to the orbitofrontal prefrontal cortices whereas the paracentral nucleus projects to the occipitofrontal and prefrontal associations areas and the central lateral nucleus projects mainly to parietotemporal association areas (Bogousslavsky and Carrera, 2005).

After paramedian infarcts involving the intralaminar nuclei (including the paracentral, central lateral, centromedian and parafascicular) but also the dorsomedian nucleus the patients, physically and emotionally active before stroke, may become severely apathetic, aspontaneous, indifferent as if they have lost motor and affectic drive, especially after bilateral lesions (Bogousslavsky et al., 1991; Engelborghs et al., 2000; Bogousslavsky and Carrera, 2005). These patients need for constant external programming, which makes them look like robots. The term of "loss of psychic self-activation" have been proposed to describe this peculiar behaviour pattern (LaPlane, 1990; Bogousslavsky and Carrera, 2005). In monkeys, intralaminar nuclei, especially the centromedian and parafascicular nuclei, play an important role in arousal and motivation with tight connections with the basal ganglia especially to the pallidum. Involvement of the striatal-ventral pallidal-frontomesial limbic loop is suggested with interruption of posterior orbital and mesial frontal cortex projections, which project to the anterior cingulate gyrus and then back to the striatum (Haber et al., 1985; Bogousslavsky and Carrera, 2005). Additionally, limbic nuclei, especially the dorsomedian nuclear groups (in addition to the anterior nuclear group with their links to the limbic structures of the cingulate gyrus and hippocampus), may play an additional role in drive and motivation (Bogousslavsky and Carrera, 2005). Akinetic mutism may also occur after bilateral thalamic lesion and be difficult to distinguish from loss of self activation. Akinetic mutism is considered when the patients, who appear to be awake and follow the examiner's eyes, fail to respond and become active after relevant stimuli. The location of thalamic lesion is quite similar in patients with loss of self activation and in those with akinetic mutism. However, involvement of intralaminar nuclei especially the centromedian and parafascicular nuclei may be more extended in patients presenting with akinetic mutism than in those with "psychic loss of psychic self-activation" (Van Domburg et al., 1996; Bogousslavsky and Carrera, 2005).

Amnesia is a frequent and prominent sign of paramedian infarcts probably resulting from a disconnection syndrome to the frontal and temporal lobes. However patients with paramedian infarcts often present with a milder amnesia than patients with a more anterior lesion involving the MMT (Bogousslavsky and Carrera, 2005).

A patient with a lesion restricted to the intralaminar and medial nuclei (and sparing the MMT, anterior thalamic nuclei and the vast majority of the dorsomedian nucleus) presented with a less severe amnesia than patients with a lesion involving anterior nuclei and/or MTT (Mennemeier et al., 1992; Bogousslavsky and Carrera, 2005). The amnesia was characterized by severe distractibility, which suggests that the intralaminar nuclei are probably not memory structures per se but may be part of a functional system regulating attention for simultaneous activities (Mennemeier et al., 1992; Bogousslavsky and Carrera, 2005).

The analysis of two cases with lesions limited to the dorsomedian nuclei (DM) (Kritchevsky et al, 1987; Bogousslavsky and Carrera, 2005) without any discernible memory deficit suggests that DM damage only causes amnesia if there is a lesion of a second thalamic structure.

Primate studies showed that the anterior thalamic thalamic nuclei are strongly related to the hippocampal formation (MTT tract) (Nauta, 1962; Graff-Radford et al., 1990; Rosene et al. 1981; Bogousslavsky and Carrera, 2005) whereas the DM nucleus has stronger connections to the amygdala (amygdalo-fugal tract) (Nauta, 1962; Graff-Radford et al., 1990; Rosene et al. 1981; Bogousslavsky and Carrera, 2005). The difference in severity of amnesia after MTT and anterior nuclei lesions compared with amnesia after DM nucleus damage may be explained by involvement of these two different limbic circuits (Nauta, 1962; Graff-Radford et al., 1990; Rosene et al. 1981; Bogousslavsky and Carrera, 2005). Damage to both limbic circuits may result in the most severe cases of amnesia (Nauta, 1962; Graff-Radford et al., 1990; Rosene et al. 1981; Bogousslavsky and Carrera, 2005).

A disinhibition syndrome may occur after paramedian infarcts, although apathy and loss of psychic self-activation are most frequently encountered. When present, the disinhibition syndrome, resulting from thalamofrontal disconnection, is often difficult to distinguish from psychiatric pathologies (Bogousslavsky and Carrera, 2005).

Different forms of beaviour disnihibition occur; a patient described (Bogousslavsky et al., 1988) showed logorrhea, delirium, joking, laughing, improper comments after right thalamic infarct involving the anterior parts of the dorsomedian and the intralaminar nuclei. She could start a conversation adequately but, immediately afterward, this was interrupted by inappropriate and fantastic stories. The patient looked like patients with medial right prefrontal lesion with speech disinhibition, "fuite des idées" and loss of selectivity of the mental processes. Interestingly, this disinhibition syndrome, almost exclusively restricted to speech, contrasted with a persistent lack of spontaneity. SPECT studies, 10 days after admission, showed a marked hypoperfusion in the right thalamic region with an associated hypoperfusion of the overlying cortex, mainly in the frontal region (Bogousslavsky et al., 1988; Bogousslavsky and Carrera, 2005).

After a bilateral paramedian infarct involving the anterior part of intralaminar nuclei and the contiguous part of the DM nucleus, a patient (McGilchrist et al., 1993; Bogousslavsky and Carrera, 2005) presented an acute-onset cyclical behaviour change. The patient, cheerful and lucid before stroke became apathetic and was "like a zombie", sleeping and eating excessively. However, every three months, he became alert, played with his children and slept little during the night. During these episodes his behaviour was frankly manic: the patient prayed, cried and had a labile affect with flight of ideas. After 36 hours he relapsed to his former apathetic state. With this report, a role of the thalamus in Kleine-Levin syndrome was suggested. SPECT scan showed severe hypoperfusion of both frontal lobes (left more marked than right) and of both thalami suggesting a thalamo-cortical disconnection (McGilchrist et al., 1993; Bogousslavsky and Carrera, 2005) .

Utilization behaviour was described after right thalamic infarct, involving anterior parts of intralaminar and dorsomedial nuclei bilaterally but also after a lesion of the same nuclei but only on the right side (Eslinger et al., 1991; Hashimoto et al., 1995; Moreaud et al., 1996; Bogousslavsky and Carrera, 2005).The patient with bilateral lesions excessively utilized objects and seemed unable to regulate her interaction with the environment. In this patient, utilization behaviour was associated with instinctive grasp reaction and motor impersistence (Eslinger et al., 1991). SPECT studies of the patient with the right thalamic lesion showed hypoperfusion of the entire right cerebral cortex with some predominance of the frontal area. The hypothesis of thalamo-cortical projections interruption is supported by the occurrence of the same behaviour pattern in a patient with an infarction of the genu of the internal capsule (Eslinger et al., 1991; Hashimoto et al., 1995; Moreaud et al., 1996; Bogousslavsky and Carrera, 2005).

A patient showed, after a bilateral dorsomedian nuclei infarcts, a childish behavior (Fukatsu et al., 1997).

A patient with persistent Klüver-Bucy syndrome was reported after bilateral lesions of dorsomedian nucleus and part of the intarlaminar nuclei (Muller et al., 1999; Bogousslavsky and Carrera, 2005). The patient, who had an active social and intellectual life before stroke, developed severe behavioral abnormalities. Two days after stroke, when consciousness disturbances resolved, she started to ate and smoke compulsively. She stopped only when food or cigarettes were out of sight. The patient also expressed a lack of shame: she walked about her room undressed, flirted, joked and spontaneously. She was highly distractible and reacted to every auditory, visual, olfactory and tactile cue, In this patient with bilateral paramedian infarct, PET studies revealed decreased 18 fluorodeoxyglucose uptake in the anterior parts of both thalami and both frontotemporal cortices. Interruption of the limbic circuitry at the site of the dorsomedian nucleus was proposed to induce Klüver-Bucy syndrome after bilateral paramedian infarcts (Bogousslavsky and Carrera, 2005).

Aphasia after paramedian infarcts is suggestive of transcortical motor aphasia with reduced fluency sparing comprehension and repetition (Kumar et al., 1996; Bogousslavsky and Carrera, 2005). Exceptionally, with extensive left thalamic lesions , comprehension and repetition are impaired.

Three main roles of thalamus have been proposed in aphasia 1) attention and speech initiation 2) lexical selection 3) motor production. Hypophonia associated with loss of initiative may be found even leading to voice extinction (Guberman and Stuss, 1983; Bogousslavsky and Carrera , 2005). Dysarthria, although less frequent than after striato-capsular infarcts may occur after paramedian infarcts and is particularly severe in case of bilateral lesions (Ackermann et al., 1983; Bogousslavsky and Carrera, 2005). Semantic paraphasias more often than phonologic paraphasias are frequently found especially when performing naming tests (Kennedy and Murdoch, 1993; Moretti et al., 2003 C).

Nadeau and Crosson (1997) distinguished two phonologic axes between anterior and posterior areas of language. They postulated that lesions to the direct axis, travelling in the arcuate fibres, induces phonologic paraphasias as seen in conduction aphasia, whereas involvement of an indirect axis travelling in the thalamus results in semantic paraphasias (Nadeau and Crosson, 1997; Bogousslavsky and Carrera, 2005).

In contrast to amnesia, thalamic aphasia seems not to depend on restricted thalamic structures. Bruyn (1989) suggested a frontal rostro-caudal/thalamic mediolateral interrelationship irrespective of thalamic nuclear masses. Nadeau and Crosson (1997) postulated a circuit responsible for thalamic aphasia which includes the frontal lobes, the nucleus reticularis, the inferior thalamic peduncle which connects the frontal lobes with the anterior portions of the nucleus reticularis and possibly the centromedian nucleus (Crosson, 1999; Bogousslavsky and Carrera, 2005).

The lateral nuclei consist mainly of the ventrolateral nucleus and the ventroposterior nuclei (including the ventroposterolateral, the ventroposteroinferior and the ventroposteromedian nuclei) but also the ventromedian nucleus. The ventrolateral nucleus has complex connexions to the motor cortex but also to the prefrontal and posterior cortices and to the upper bank of the superior temporal cortices (Bogousslavsky and Carrera, 2005). The ventroposterolateral nucleus receives the medial lemniscal and spinothalamic pathways (body and limbs sensibility) and the ventroposteromedial nucleus is the relay of the trigeminothalamic pathway (face and neck). The ventroposterolateral and the ventroposteromedian nuclei have reciprocal connections to the primary sensory cortex (Strick, 1976; Bogousslavsky and Carrera, 2005). A mediobasal part of the ventroposteromedial nucleus receives projections of taste fibres (Bogousslavski and Carrera, 2005). The ventroposterolateral nucleus has connections with the vestibular nucleiand receives projections from lemnsical fibers (Bogousslavsky and Carrera, 2005).

The most common clinical pattern after inferolateral infarct is a controlateral ataxia and hypesthesia. Mild paresis may be found in the acute found but is in most cases transient. Neuropsychological and neurobehavioral disorders more rarely occur after inferolateral infarcts, especially after left-sided lesions. In a recent report (Weisman et al., 2004; Bogousslavsky and Carrera, 2005), a patient with left inferolateral infarct showed conduction aphasia with severe impairment of repetition.

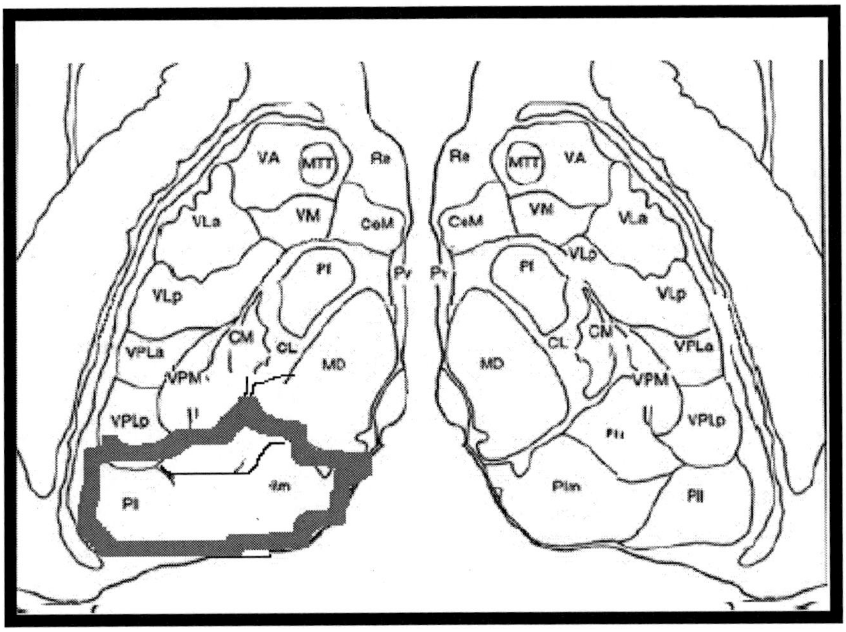


Figure 27. A cartoon for an ideal representation of the inferolateral thalamic nuclei (modified by the Authors).

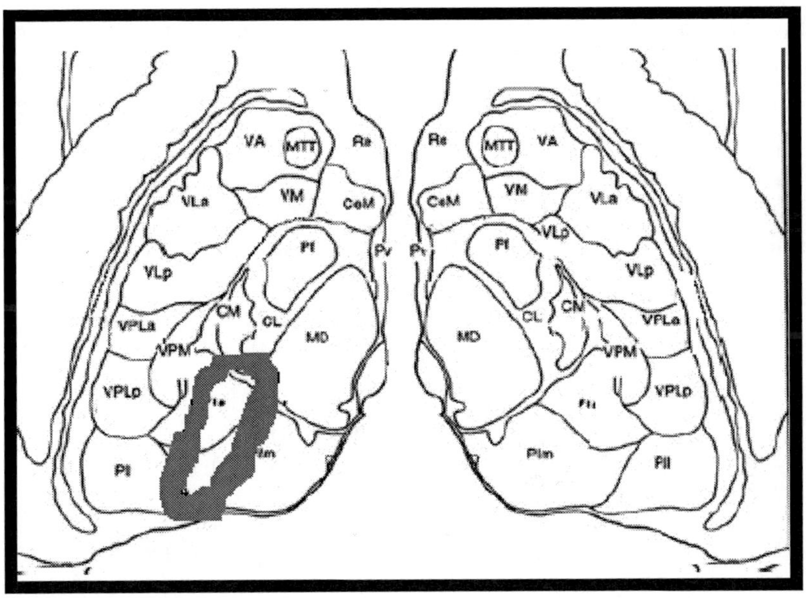


Figure 28. A cartoon for an ideal representation of the posterior thalamic nuclei (modified by the Authors).

Interruption of corticothalamo- cortical tracts disconnecting temporal and frontal lobe was suggested (Weisman et al., 2004; Bogousslavsky and Carrera, 2005).

The pulvinar is the main component of posterior nuclei. It can be divided into three main parts : the medial pulvinar, which projects mainly to the parietooccipital cortex, the prefrontal cortex, the paralimbic cingulated and parahippocampal cortices (Bogousslavsky and Carrera, 2005). The inferior and lateral pulvinar, which have reciprocal connections to the striate and extrastriate cortex of the occipital lobe, parietal and temporal cortices (Bogousslavsky and Carrera, 2005).

Infarcts in the territory of the posterior choroidal artery is rare. Pulvinar is mainly supplied by two subdivisions

of the posterior choroidal artery (Neau and Bogousslavsky, 1996): the medial and the lateral arteries. The most frequent sign are controlateral hypesthesia and homonymous horizontal sectoranopsia (Neau and Bogousslavsky, 1996) . Neuropsychological signs include neglect and aphasia. Patients with posterior chordial artery infarct may show severe retrograde and anterograde amnesia, faulty retention of verbal material and form discrimination and amnestic dysnomia probably because of simultaneous hippocampus and/or parahippocampal gyrus involvement (Neau and Bogousslavsky, 1996; Bogousslavsky and Carrera, 2005).

Controlateral motor neglect is reported after right pulvinar nucleus (Neau and Bogousslavsky, 1996; Bogousslavsky and Carrera, 2005). It has been suggested that centromedian and parafascicular nuclei are closely associated with motor systems (basal ganglia, ventrolateral nucleus of the thalamus and frontal lobes) (Watson et al., 1981; Bogousslavsky and Carrera, 2005). Controlateral spacial neglect is related to perception of space and is found after anterior, median, lateral and posterior nuclei damages (Watson et al., 1981; Bogousslavsky and Carrera, 2005). However, a recent study (Karnath et al., 2002) based on 7 patients showed that pulvinar, especially its medial part, is the most frequent thalamic region associated with spatial neglect probably resulting from interruption of thalamo-projection to the ipsilateral temporal rather than parietal lobe (Karanath et al, 2001; Karnath et al., 2002; Bogousslavsky and Carrera, 2005). Neglect is more frequent after right lesion but also occurs after left lesion (Motomura et al., 1986; Bogousslavsky and Carrera, 2005).

In: Basal Ganglia and Thalamus
Editors: R. Moretti, P. Torre and R. M. Antonello
ISBN 978-1-60692-198-2


*Chapter 17*

# Imaging of Basal Ganglia

***Maja Ukmar and Elena Trincia***
Dipartimento di Radiodiologia , Ospedale Cattinara Hospital,
Università di Trieste, Italy

The technique of choice in the evaluation of the basal ganglia is magnetic resonance imaging (MRI) although computed tomography (CT) is to date better in the detection of calcifications.

On CT images they appear rather hyperdense to the white matter and isodense to the cortex. (Figure 29). They frequently show some punctate hyperdensity due to the deposition of calcium, especially in the middle aged or old patients.

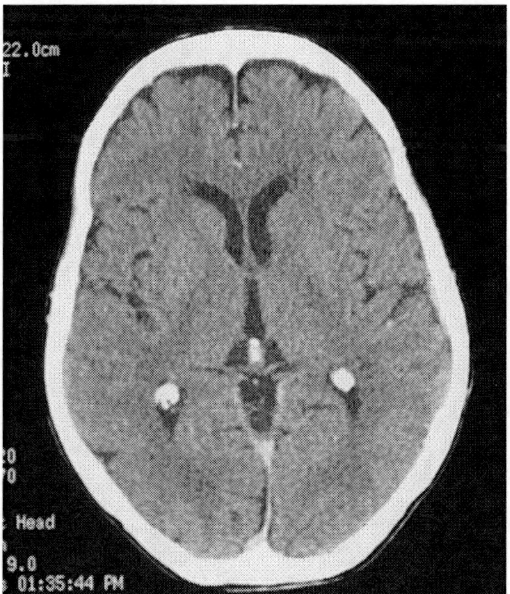


Figure 29. A CT image of normal appearing of basal ganglia.

Considering MRI the basal ganglia should be evaluated by using high field magnets and spin-echo or GE sequences which are more sensitive in the detection of iron deposition in the brain. On the T1 weighted images and fluid attenuation inversion recovery (FLAIR) they appears isointense to the cortex and slightly hypointense to the surrounding gray matter. (Figure 30)

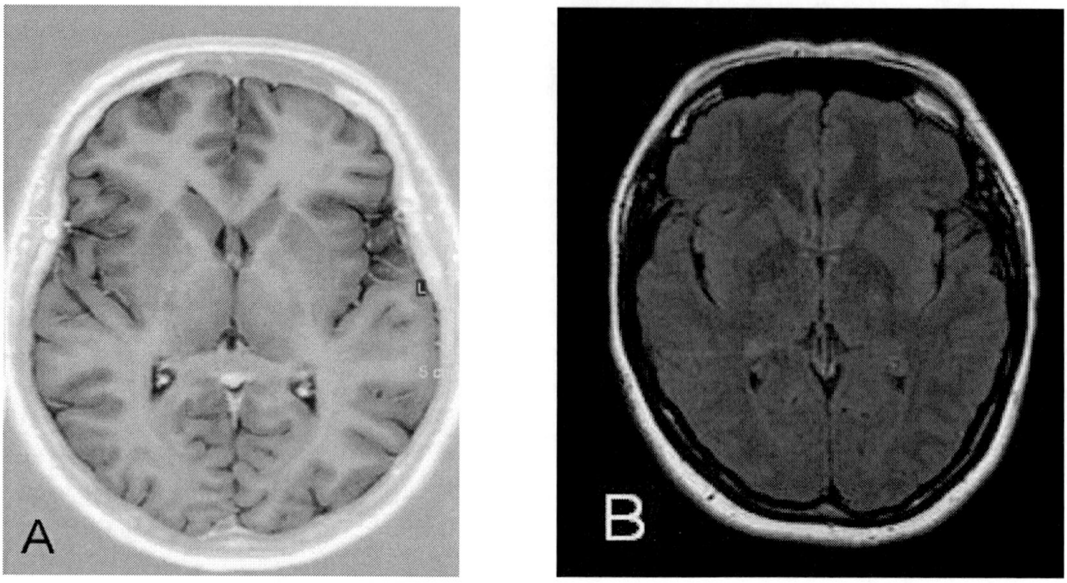


Figure 30. MR images of normal appearing basal ganglia. A) T1 IR weighted image, B) FLAIR weighted image

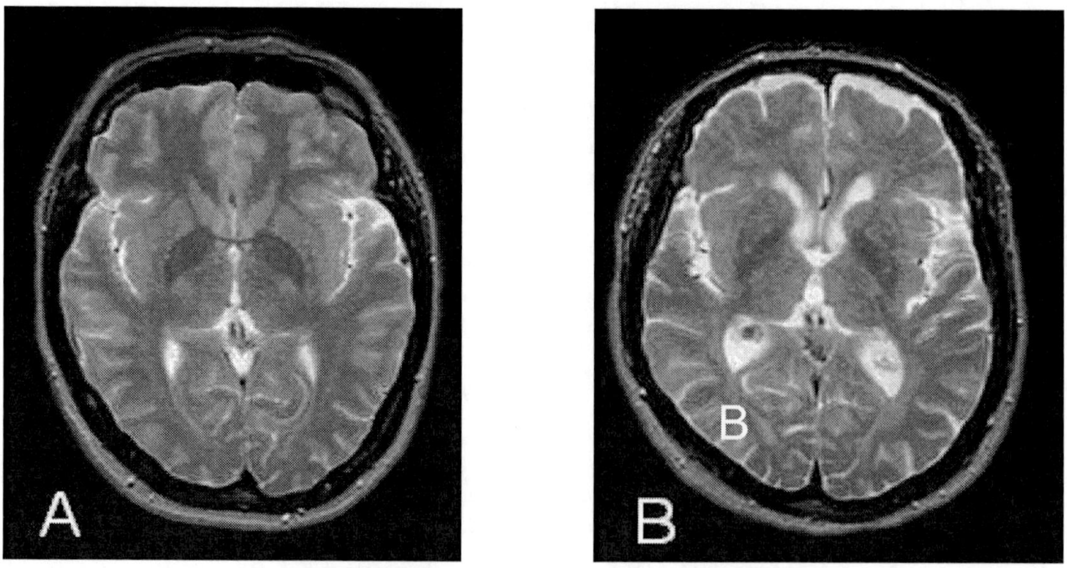


Figure 31. MR images of normal appearing basal ganglia. A) T2 weighted image in a young patient. B) T2 weighted image in normal elderly patient.

On T2 weighted images their appearance changes with aging and it is correlated with the presence of iron which is absent at birth in the brain. With the increase of age there is a progressive decrease of the intensity of signal on T2 weighed sequences. The last nuclei to became hypointense is putamen, usually in patients older than 70 years (Milton et al. 1991). On T2 weighted images is rather common to find focal or diffuse hyperintensities in normal elderly patients which correspond to arteriosclerotic demyelination.(Figure 31)

Many diseases with different etiology involve the basal ganglia, they can be classified in degenerative, inflammatory- infective and toxic-metabolic disorders.

In these chapter we shortly examine the most common disorders for each class.

# Degenerative Disorders

## Parkinson's Disease

Both CT and MRI are not useful in the evaluation of idiopathic Parkinson's disease (PD). Most frequently the imaging is normal or shows non specific atrophy. Some authors describe on MR T2 images narrowing of the pars compacta of substantia nigra and focal hyperintensity or hypointensity in the substantia nigra (Dagher, 2001) (Fig 32). The underlying pathology of these lesions is loss of dopaminergic cells and gliosis. In some patients hyperintensities in the putamen were found overlapping those found in healthy elderly patients.

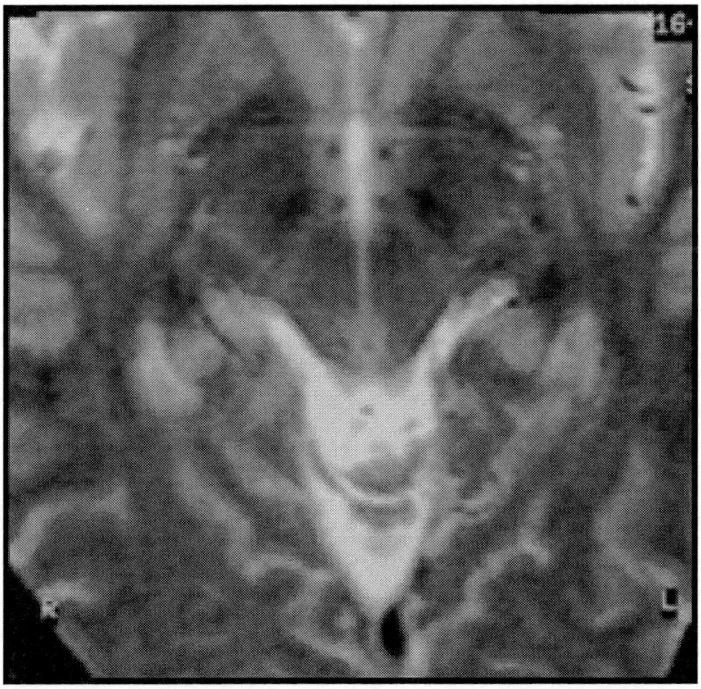

Figure 32. T2 wieghted images in PD patient. The image shows narrowing of the pars compacta of substantia nigra. Multiple system atrophy (striatonigral degeneration).

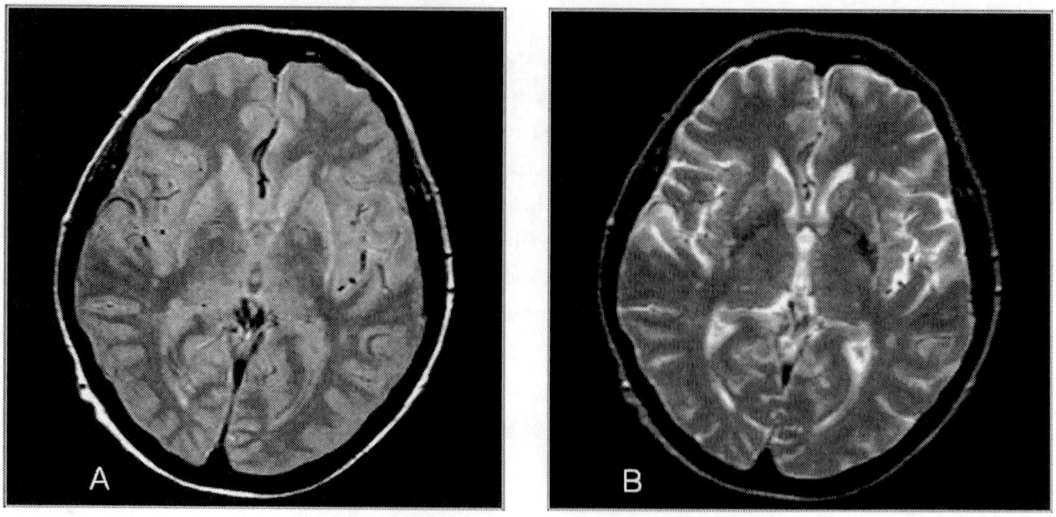


Figure 33. A-B: PD and T2 weighted images in MSA patients. The images show putaminal atrophy in particular on the left associated to hypointensity in the dorslateral part of the putamen bilaterally and hyperintense rim on the outer margin of the nuclei.

CT is not useful in the evaluation of patients with multisystemic atrophy (MSA). The atrophy of putamen is better shown on MR images. Common finding on T2 weighted images is a hypointensity of the dorsolateral putamen due to iron deposition, although these findings are not specific for MSA. More specific sign is represented by a hyperintense rim, better seen on proton density (PD) weighted images on the outer margin of the putamen, probably consequence of the trans-sinaptic degeneration of the striato-nigral system (Fig 33). The signs could be absent in the early stage of the disease (Schrag et al., 2000).

# Infectious Disease

## Creutzfeld-Jakob Disease

Creutzfeld-Jakob disease (CJD) is the most common of so called transmissible spongiform encephalopathies. From the pathological point of view is characterized by spongiform degeneration of neurons in the cortex and in the subcortical nuclei. The classical form has a sporadic origin whereas the CJD variant is transmitted from animals.

CT could be normal in the early stage of the disease. With progression of symptoms can show non specific atrophy.

MRI is the technique of choice in the evaluation of CJD. On the T2 weighted and FLAIR images the caudate nucleus and putamen show bilaterally high intense signal intensity. More recently diffusion imaging demonstrated earlier hyperintensity of the basal ganglia. In the variant form, but in some cases also in the sporadic one, a hyperintensity of pulvinar of the thalami can occur. In the late stages atrophy and white matter hyperintensities develop ( Figure 34) (Collie et al, 2003).

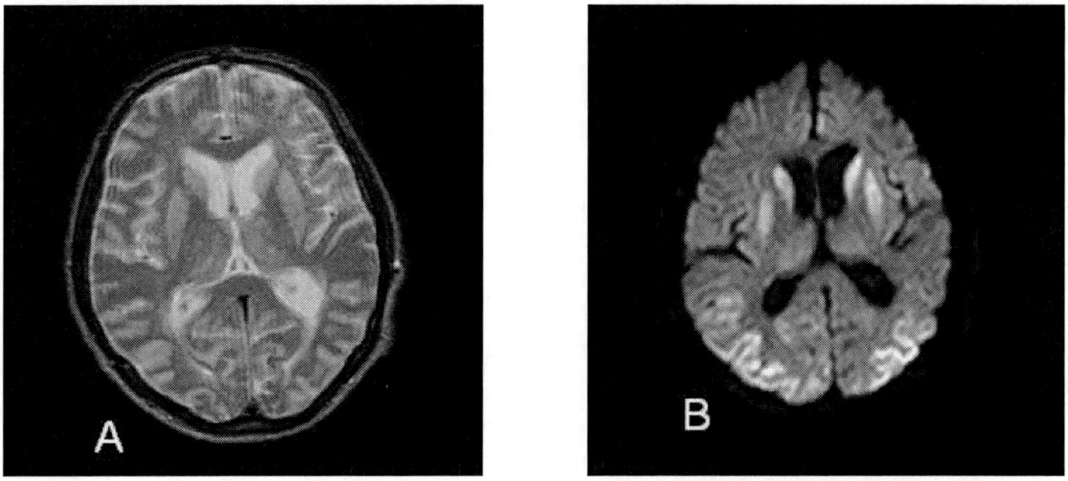


Figure 34. MR images of patient with CJD. A) On T2 weighted image bilateral and symmetrical hyperintensity in the basal ganglia. B) On DWI cortical hyperintensity is better shown.

# Metabolic- Degenerative Inherited Disorders

## Wilson Disease:

Wilson disease (WD) is an inborn error of copper metabolism due of ATP7B mutation localized to chromosome 13. It is an autosomal recessive disease which comport a defective excretion of copper from hepatocytes, resulting in excessive accumulation in the liver, brain in particular with a sofference and degeneration of basal ganglia (Das et al., 2007). Widespread copper accumulation in central nervous system results in cognitive, extrapyramidal and pyramidal dysfunction.

CT is often normal, copper depositions does not increase the density and not assume contrast-agent. Sometime a cortical and cerebral atrophy can be evaluated.

MRI is surely more sensitive than CT. MRI can shows abnormalities even in the absence of clinical neurological findings and reveals a signal alteration in basal ganglia, in order in the putamen, globus pallidus, caudate and thalami. Basal ganglia lesions are rather symmetrical, whereas cortical and subcortical lesions are non symmetric.

On MRI images the most frequent finding is a concentric-laminar increase in signal intensity on T2-weigheted sequences . Sometimes a striking outer rim of high intensity on T2-weighted images and a diffuse involvement can be evaluated.

In T2-weigheted images and FLAIR a characteristic " face of panda" sign could be observed on axial plane at midbrain level which is characterized by a hyperintensity in tegmentum, a hypointensity of superior colliculus and preserved signal intensity of lateral portion of pars reticulate of substantia nigra.

T1-w sequences show a hypointensity, which reflects the underlying edema, gliosis and cavitation (King et al., 1996).

### Leigh Syndrome:

Leigh Syndrome (LS) is a progressive neurodegenerative disorder which usually onset in infancy or early childhood. It is a genetically heterogeneous mithochondrial disorder which evolve to a progressive neurodegeneration characterized by psychomotor retardation, feeding difficulties, intermittent abnormalities of the respiratory rhythm and ataxia (Arii et al., 2000).

Symmetrical spongiforms lesions are the typical neuropathologic findings which are the result of a neuropil vacuolation and a relative preservation of the neuronal cells with a capillary proliferation. These symmetrical lesions frequently occur in the basal ganglia, in particular in the putamen and globus pallidus, and in brain stem especially in the peri-aqueductal gray matter.

CT imaging could show symmetrical hypodense lesions in the basal nuclei. Contrast-enhancement CT could reveals an uncommon enhancement in this location.

The most radiological common and typical finding is demonstrated on T2-weighed and FLAIR sequences, in particular a bilateral and symmetric focal hypeintensity. These lesions are often characterized by a hypointesity on T1- weighed imaging and after administration of contrast-agent they have an uncommon enhancement, as in CT.

Conventional MRI can fail to detect early lesions, otherwise diffusion- weighed MR imaging could show a restrictive diffusion in the acute and early stages(Arii et al., 2000). Magnetic resonance spectroscopy (MRS) can demonstrate a decreased NAA (N-acetyl aspartate) in the basal ganglia and an increase of choline as other kinds of neuronal damages.

## Toxic and Metabolic Acquired Disorders

### Methanol Intoxication:

Acute methanol poisoning is a rare accidental or suicidal intoxication, which produces a severe metabolic acidosis and serious neurologic sequelae. The peculiarity of methanol poisoning is the latency period between its ingestion inhalation or transdermal exposure and the appearance of manifestations (Sefidbakht et al. 2007). Putaminal necrosis, with or without hemorrhage, and subcortical white matter lesions are the most frequent reported findings (Blanco et al., 2006). The mechanism responsible for selective putaminal necrosis is unknown. It has been postulated that it results from decreased blood flow through the basal veins of Rosenthal due to the hypotension (Sefidbakht et al. 2007). Another possibility is that the necrosis occurs as a direct toxic effect of formic acid which has a more accumulation in the putamen than in other areas of the brain (Sefidbakht et al. 2007). Putaminal changes on conventional CT and MRI scans are not specific to methanol intoxication and have been described in various neurodegenerative disorders or in Wilson's disease or Leigh disease.

Many times cranial CT can be normal especially on the admission time. In a second time CT can shows a diffuse white matter hypodensity and putamina with acute toxic edema associated to high attenuation foci due to hemorrhage. Meningeal and gyral enhancement is reported after the administration of contras agent.

MRI gives better anatomical detail and reveal some small hemorrhagic lesions (Sefidbakht et al. 2007). On T2 W- images and FLAIR bilateral increased signal in the basal nuclei, especially in the putamina, with reciprocal hyponitensities on T1-weighted image, is the most common findings of methanol poisoning. This hyperintensity can extend also to the corona radiata, centrum semiovale and deep white matter.

In same cases in the subacute timing an atrophy of optic chiasm and prechiasmatic optic nerves have been demonstrated.

Diffusion- weighted imaging is show bilateral putaminal hyperintensity with a decreased apparent diffusion coefficient (ADC) values.

In significant methanol intoxication it is possible to find out some enhancing lesions in the caudate nuclei, putamina, hypothalamus and subcortical white matter after the administration of Gadolinium contrast agent (Sefidbakht et al. 2007).

## Acute Diabetic Uraemia:

Symmetrical bilateral basal ganglia lesions my be caused also by severe azotemia or end-stage diabetic nephropathy. The precise pathogenesis is still unknown, even if the high frequency microvascular changes in the diabetic patients could increase susceptibility of basal ganglia to insults and suggest a possible ischaemic aetiology.

On the other hand uraemic toxins have been implicated in the pathogenesis of these lesions in the basal ganglia, which are particularly vulnerable to a wide range of toxins and metabolic disturbances (Li et al., 2008).

In these kinds of patient CT and MRI have shown bilateral basal ganglia lesions which, in contrast to the other metabolic disorders of basal ganglia, regress spontaneously .

CT shows bilateral basal ganglia isolated hypodensity involving the lentiform and caudate nuclei associated with mild surrounding, but no focus of haemorrhage (Li et al., 2008). On MRI T1 weighted sequences reveal a symmetric decrease of signal intensity . While T2–w images and FLAIR are characterized by hyperintense areas, without haemorrhagic foci. Diffusion imaging demonstrates patchy areas of high signal in the basal ganglia, most strikingly in the lentiform nuclei. ADC map shows a low signal, indicating intracellular oedema.

A follow-up imaging show almost complete regression of bilateral basal ganglia lesions with minimal surrounding oedema (Oh et al., 2002; Li et al., 2008).

## Hyperglycemia:

Acute generalised chorea can be attributed to multiple causes, including non-ketotic hyperglycemia and it occurs predominantly in women.

CT shows a spontaneous bilateral hyperdensity in the putamen and caudate nuclei.

On T1 weighted sequence characteristic high signal intensity lesion my be seen in the basal ganglia and in particular the putamen is always involved (Oh et al., 2002). The high signal intensity basal ganglia lesion on T1-w images is reversible and correlated with the clinical improvement in chorea.

In contrast to the hyperintensity on T1-w sequences , the T2-w studies show iso-hypointensity lesions in the same foci. In same cases, calcium deposits and focal microhaemorrahage are found in the lesioned putamen and caudate (Mestre et al., 2006).

## Fahr Disease:

It is known also as cerebrovascular ferrocalcinosis or Idiophatic basal ganglia calcification (IBGC). Fahr disease a rare degenerative neurological disorder characterized by extensive capillary an parenchymal bilateral deposition of calcium in the basal nuclei. Otherwise prominent calcifications may also occur in the dentate nuclei, centrum semiovale and subcortical white matter.

CT is more sensitive than MRI in the evaluation of calcification and reveals focal areas of increased attenuation in the basal ganglia in particular in the globus pallidus (Figure 35) In MR studies with T1-w sequences the signal intensities calcified lesions have various appearance, depending on the stage of the disease and the volume of calcium deposits, otherwise they are usually hyperintense. On T2- w images some hyperintense areas are visible which does not correspond to the deposits of calcium, but they can reflect some metabolic or inflammatory process ( Weisman et al., 2007).

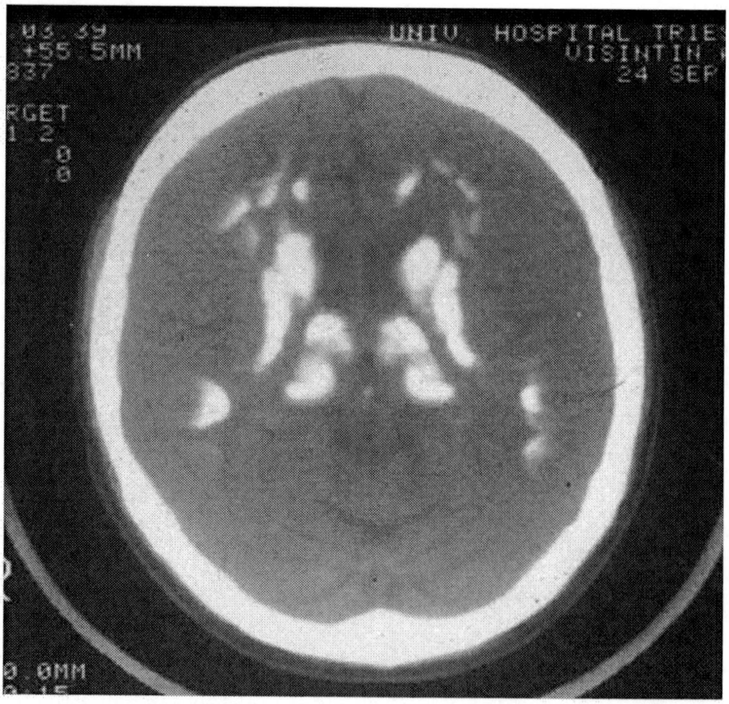


Figure 35. A CT image in patient with Fahr disease: bilateral calcification of the basal ganglia and subcortical white matter.

## Carbon Monoxide Intoxication:

Carbon monoxide (CO) is a toxic gas produced by an incomplete combustion of hydrocarbons. CO affects the cytochrome oxidase system and it causes an anoxic-ischemic encephalopathy in particular in the most sensitive brain regions as the cerebral cortex and the basal ganglia, in particular caudate nuclei and putamina. In fact regions with relatively poor vascularization, such globus pallidus, are more vulnerable, in particular during periods of hypotension. The typical pallidum lesions are well- defined, with a bilateral macroscopic infarctions. Sometimes a small linear focus of necrosis is found at the junction of the internal capsule and the internal nucleus of the globus pallidus.

On CT imaging a symmetric hypodensity in the globus pallidum has been reported, which in some cases could include the white matter.

MR imaging is more sensitive than CT (O'Donell et al., 2000). T2 and FLAIR weighted images are more diagnostic than T1-w ones. T2-w sequences show bilateral hyperintensity in the globus pallidus, often surrounded by hypointense rim. DWI shows a restricted water diffusion pattern in globus pallidus and substantia nigra. Magnetic resonance spectroscopy (MRS) can demonstrate a decreased NAA (N-acetyl aspartate) in the basal ganglia, as a result of brain damage. In early CO-intoxication Proton MRS (H-MRS) shows a persistent increases in choline related to progressive demylination. In irreversibly damage lactate appears and NAA decreases (Prockop and Chichkova, 2007).

## Hepatic Encephalopathy:

Hepatic encephalopathy (HE) is a neurological disturbance which complicates the acute or chronic liver disease. This syndrome is usually considered a metabolic derangement of brain functions for toxins accumulating in the blood and in the brain, in particular in the cortex and basal ganglia. Microscopically there is astocytosis and sometimes loss of neurons and fibres like demylination (Zeneroli,1991). In acute HE, CT can show a severe and diffuse cerebral edema. In chronic HE, a cerebral atrophy is frequently observed. No abnormalities can be appreciated in the basal ganglia.

On MR imaging a peculiar finding is a bilateral and symmetrical hyperintensity of the globus pallidus in the T1-weigheted images in its inner and outer margin. It corresponds to shortening of T1-relaxion time due to the accumulation of paramagnetic compounds as manganese or free radical. This hypentensity of the globus pallidus seems to be linked with the duration of HE. In T2-w sequences usually no signal alterations can be evaluated in the basal ganglia. On the other hand cortex and periventicular matter can appear hyperintense in T2-w sequences (Zeneroli, 1991; Klos et al., 2006).

In: Basal Ganglia and Thalamus
Editors: R. Moretti, P. Torre and R. M. Antonello
ISBN 978-1-60692-198-2


# Conclusions

***Rita Moretti, Paola Torre, and Rodolfo M. Antonello***

The properties of neurons within the basal ganglia resemble the properties of neurons within the cortical areas subserved by the specific loop; in so far, neuronal activity within basal ganglia and cerebellar loops with areas of the motor cortex is highly correlated with motor parameters, and damage to these neurons leads to motor symptoms; neuronal activity within basal ganglia with areas of the prefrontal cortex is more related to voluntary deliberation, motivation and complex perception of the self, and the disruption of these circuits lead to higher –order deficits.

Though these aspects, every neurologist or neuropsychologist who faces with the clinical variety of patients, admits that basal ganglia or thalamus lesions cause a panoply of complex symptoms, signs and even syndromes. The myriad of symptoms described in basal ganglia pathologies seem to be determined by the maintenance of many of the physiological and behavioural properties of the cortical areas that they subserve, but also by the possible influence of reward on neurons within prefrontal-basal ganglia circuits, by their strenuous connections with different brain areas and by their intimate relationships.

We have tempted to collect many data from Literature and from clinical practice, remembering that subjectivity leads the neurology not to be an exact science but a most admirable form of art. Our attempt has been to respect the art, by looking at the scientific data.

In: Basal Ganglia and Thalamus
Editors: R. Moretti, P. Torre and R. M. Antonello
ISBN 978-1-60692-198-2


# References

Aarsland D, Larsen JP, Lim NG, Janvin C, Karlsen K, Tandberg E, Cummings JL. Range of neuropsychiatric disturbances in patients with Parkinson's disease. *J. Neurol. Neurosurg. Psychiatry*. 1999; 67:492--496.

Aarsland D, Litvan I, Larsen JP. Neuropsychiatric symptoms of patients with progressive supranuclear palsy and Parkinson's disease. *J. Neuropsychiatry Clin. Neurosci.* 2001;13:42--49.

Abbott RD, Petrovitch H, White LR, Masaki KH, Tanner CM, Curb JD, et al. Frequency of bowel movements and the future risk of Parkinson's disease.[comment]. *Neurology.* 2001;57(3):456-462.

Abdullaev Y.G., Bechtereva N.P., Melnichuk, K.V. Neuronal activity of human caudate nucleus and prefrontal cortex in cognitive tasks. Behav. Brain Res. 1998; 97: 159-177.

Ackermann H, Ziegler W, Petersen D. Dysarthria in bilateral thalamic infarction. A case study. *J. Neurol.* 1983;240:357-362.

Adams NC, Lozsadi DA, Guillery RW. Complexities in the thalamocortical and corticothalamic pathways. *Eur. J. Neurosci* . 1997; 9:204–209

Agid Y, Graybiel AM, Ruberg M, Hirsch E, Blin J, Dubois B, et al. The efficacy of levodopa treatment declines in the course of Parkinson's disease: do nondopaminergic lesions play a role? In: Streifler MB, Korczyn AD, Melamed E, Youdim MB, editors. Parkinson's Disease: Anatomy, Pathology, and Therapy. New York: Raven Press; 1990. p. 83-100.

Aglioti S., Beltramello A., Girandi F., Fabbro F. Neurolinguistic and follow-up study of an unusual pattern of recovery from bilingual subcortical aphasia. *Brain* 1996; 119: 1551-1564.

Aglioti S., Fabbro F. Paradoxical selective recovery in a bilingual aphasic following subcortical lesions. *Neuroreport*, 1993; 4: 1359-1362.

Aharon-Peretz J., Kliot D., Tomer R. Behavioral differences between white matter lacunar domentia and Alzheimer's disease: A comparison of the neuropsychiatric inventory. *Dement. Geriatr. Cogn. Disord*. 2000; 11: 294-298.

Ahl B, Weissenborn K, van den Hoff J, Fischer-Wasels D, Köstler H, Hecker H, Burchert W. Regional differences in cerebral blood flow and cerebral ammonia metabolism in patients with cirrhosis. *Hepatology.* 2004 Jul;40(1):73-9

Ahlskog JE. Challenging conventional wisdom: The etiologic role of dopamine oxidative stress in Parkinson's disease. *Movement Disorders* 2005;20.

Akopyan NS, Baklavadzhyan OG, Sarkisyan NV. The effects of the mediodorsal nucleus of the thalamus on respiratory neurons of the medulla oblongata and respiration in rats in conditions of hypoxia. *Neurosci. Behav. Physiol.* 2000; 30:449–453

Albin RL, Young AB, Penney JB. The functional anatomy of basal ganglia disorders. *Trends Neurosci.* 1989;12:366–375.

Alexander G. A. Basal ganglia thalamocortical circuits: their role in control of movement. *Journal of Clinical Neurophysiolgy* 1994; 11: 420-431.

Alexander G.E., DeLong M.R, Strick P.L Parallel organization of functionally segregated circuits linking basal ganglia and cortex. *Annu. Rev. Neurosci.*1986; 9, 357-381.

Alexander M. Clinical- anatomical correlations of aphasia following predominantely subcortical lesions. In F. Boller and J. Grafman (Eds.), Handbook of neuropsychology, vol. 2. Amsterdam: *Elsevier.* 1989. Pp. 47-66.

Alexander M., Naeser P., Palumbo C. Correlations of subcortical CT lesion sites and aphasia profiles. *Brain*, 1987; 110: 961-991.

Alexopoulos GS., Abrams RC., Young RC., Shamoian CA. Cornell Scale for Depression in *Dementia. Biol. Psychiatr.* 1988; 23: 271-284.

Alheid G.F., Heimeri L, Switzer R.C. Basal ganglia, in The Human Nervous System, Academic Press, New York,1990. Pp. 483-582.

Almor A, Kempler D, Andersen ES, McDonald MC, Hayes UL, Hintiryan H. The production of regularly and irregularly inflected nouns and verbs in Alzheimer's and Parkinson's patients. *Brain Lang* 2002; 83: 149–51.

AMERICAN PSYCHIATRIC ASSOCIATION. Diagnostic and statistical manual of mental disorders: DSM-IV (4th ed.) Washington, DC: American Psychiatric Association 1994.

Ardouin C., Pillon B., Peiffer E., Bejjani P., Limousin P., Damier P., Arnulf I., Benabid, A.L., Agid Y., Pollak P. Bilateral subthalamic or pallidal stimulation for Parkinson's Disease affects neither memory nor executive functions: a consecutive series of 62 patients. *Annals of Neurology* 1999; 46(2): 217-223.

Arii J, Tanabe Y. Leigh syndrome: serial MR imaging and clinical follow-up. *AJNR Am. J. Neuroradiol.* 2000 Sep;21(8):1502-9.

Ashburner, J., Friston, K.J., Voxel-based morphometry—The methods. *NeuroImage* 2000; 11: 805–821.

Ashby P. What does stimulation in the brain actually do? *Prog. Neurol. Surg.,* 2000, 15, 236-245.

Baddeley A. Recent developments in working memory. *Curr. Opin. Neurobiol.* 1998; 8:234-238.

Bal T., Debay D., Destexhe A. Cortical feedback controls the frequency and synchrony of oscillation in the visual thalamus. *J. Neurosci.* 2000; 20:7478-7488.

Bang SJ, Choi SH, Park NH, Shin JW, Choy N, Kim SH, Kim Y. High pallidal T1 signal is rarely observed in obstructive jaundice, but is frequently observed in liver cirrhosis. *J. Occup. Health.* 2007 Jul;49(4):268-72

Barbas H, Haswell H, Dermon CR. Diverse thalamic projections to the prefrontal cortex in the rhesus monkey. *J. Comp. Neurol.* 1991;313:65–94.

Barbas H. Connections underlying the synthesis of cognition, memory, and emotion in primate prefrontal cortices. *Brain Res. Bull.* 2000; 52:319–330

Baron JC. Consequences of lesions of the basal ganglia for cerebral metabolic activity: clinical implications. *Rev. Neurol.* (Paris). 1994; 150:599--604.

Bartha R, Stein MB, Williamson PC, Drost DJ, Neufeld RW, Carr TJ, et al. A short echo 1H spectroscopy and volumetric MRI study of the corpus striatum in patients with obsessive-compulsive disorder and comparison subjects. *Am. J. Psychiatry* 1998;155(11):1584-91.

Basser, P.J., Pierpaoli, C. Microstructural and physiological features of tissues elucidated by quantitative-diffusion-tensor MRI. *J. Magn. Reson.,* 1996; B 111: 209–219.

Basso, A., Della-Sala, S., Farabola, M. Aphasia arising from purely deep lesions. *Cortex* 1987; 23: 29–44.

Baxter LR Jr, Schwartz JM, Bergman KS, Szuba MP, Guze BH, Mazziotta JC, et al. Caudate glucose metabolic rate changes with both drug and behavior therapy for obsessive-compulsive disorder. *Arch. Gen. Psychiatry* 1992;49:681-689

Bechara A, Damasio AR, Damasio H, Anderson SW. Insensitivity to future consequences following damage to human prefrontal cortex. *Cognition.* 1994; 50:7--15.

Benson DF, Djenderedjian A, Miller BL, et al. Neural basis of confabulation. *Neurology* 1996;1239-1243.

Benton AL., Hamshe K., Varney N., Spreen O. Contributions to neuropsychological assessment. New York: Oxford University Press, 1983.

Bergeron M, Reader TA, Layrargues GP, Butterworth RF. Monoamines and metabolites in autopsied brain tissue from cirrhotic patients with hepatic encephalopathy. *Neurochem. Res.* 1989 Sep;14(9):853-9

Bergeron M, Swain MS, Reader TA, Butterworth RF. Regional alterations of dopamine and its metabolites in rat brain following portacaval anastomosis. *Neurochem. Res.* 1995 Jan;20(1):79-86.

Bhatia KP, Marsden CD The behavioural and motor consequences of focal lesions of the basal ganglia in man. *Brain* 1994; 117:859--876.

Blanco M., Casado R., Vázquez F. Pumar J.M.CT and MR Imaging Findings in Methanol Intoxication *American Journal of Neuroradiology* 2006; 27:452-454,

Blomqvist A, Ericson AC, Craig AD, Broman J. Evidence for glutamate as a neurotransmitter in spinothalamic tract terminals in the posterior region of owl monkeys. *Exp. Brain. Res.* 1996; 108:33–44

Bodis-Wollner I. Visual and visuo-cognitive dysfunctions in Parkinson's disease. American Academy of Neurology, 2003; Minneapolis. 2DS.001 pdf. 1-8.

Bodis-Wollner. Visual deficits related to dopamine deficiency in experimental animals and Parkinson's disease patients. *Trends Neurosci.* 1990: 13:296-302.

Bogousslavsky J, Carrera E. Structure function correlations I behaviourla neurology. The thalamus. American Academy of Neurology, 2005, Minneapolis; pdf. 1-16.

Bogousslavsky J, Carrera E. Structure function correlations in behavioural neurology. The thalamus. American Academy of Neurology (AAN) Acts of the Congressa. San Francisco, 2005. 1-16.

Bogousslavsky J, Ferrazzini M, Regli F, et al. Manic delirium and frontal-like syndrome with paramedian infarction of the right thalamus. *J. Neuro. Neurosurg. Psychiatry* 1988;51:116-119.

Bogousslavsky J, Regli F, Assal G. The syndrome of unilateral tuberothalamic artery territory infarction. *Stroke* 1986;17:434-441.

Bogousslavsky J, Regli F, Delaloye B, et al. Loss of psychic self activaation with bithalamic infarction. *Acta Neurol. Scand.* 1991;83:309-316.

Bogousslavsky J, Regli F, Uske A. Thalamic infarcts: Clinical syndromes, etiology, and prognosis. *Neurology* 1988;38:837-847.

Bogousslavsky J, Van Melle G, Regli F. The Lausanne Stroke Registry: analysis of 1,000 consecutive patients with first stroke. *Stroke* 1988;19:1083-1092.B

Booth JR, Wood L, Lu D, Houk JC, Bitan T. The role of the basal ganglia and cerebellum in language processing. *Brain Res.* 2007; 1133(1):136-44.

Braak H, Del Tredici K, Rub U, de Vos RA, Jansen Steur EN, Braak E. Staging of brain pathology related to sporadic Parkinson's disease. *Neurobiology of Aging.* 2003;24(2):197-211.

Braak H, Ghebremedhin E, Rub U, Bratzke H, Del Tredici K. Stages in the development of Parkinson's disease-related pathology. *Cell Tissue Res.* 2004;318:121-134.

Bremner JD, Randall P, Scott T M, et al. MRI-based measurement of hippocampal volume in patients with combat-related posttraumatic stress disorder. *Am. J. Psychiatry* 1995;152:973-981.

Bremner JD. Functional neuroanatomical correlates of traumatic stress revisited 7 years later, this time with data. *Psychopharmacol. Bull.* 2003;37: 6-25.

Briggs G., Nebes R. Patterns of hand preference in a student population. *Cortex* 1975; 11: 230-238.

Bronte-Stewart H Deep brain stimulation for movement disorders- general principles. American Academy of Neurology, Minneapolis, 2003. (Ac.003 pdf. 1-23.

Brown RG, Pluck G Negative symptoms: the 'pathology' of motivation and goal-directed behaviour. *Trends Neurosci.* 2000; 23:412--417.

Brunner R., Kornhuber H., Seenmuller E., Suger G., Wallesch C. Basal ganglia partecipation in language pathology. *Brain and Language* 1982; 16: 281-299.

Bruyn RPM. Thalamic aphasia. A conceptional critique. J Neurol 1989;236:21-25.

Burkhard PR, Delavelle J, Du Pasquier R, Spahr L. Chronic parkinsonism associated with cirrhosis: a distinct subset of acquired hepatocerebral degeneration. *Arch. Neurol.* 2003 Apr;60(4):521-8

Canavero S., Fontanella M. Behavioral-attentional syndrome following bilateral caudate head ischaemia. *Journal of Neurology* 1998; 245: 322-324.

Cannatà AP., Alberoni M., Franceschi M., Mariani C. Frontal impairment in subcortical ischemic vascular dementia in comparison to Alzheimer's Disease. *Dementia Geriatric Cognitive Disorders* 2002; 13: 101-111.

Cappa S., Cavallotti G., Guidotti M., Papagno C., Vignolo L. Subcortical aphasia: two clinical CT scan correlation studies. *Cortex,* 1983; 19: 227-241.

Carmichael ST, Price JL. Sensory and premotor connections of the orbital and medial prefrontal cortex of macaque monkeys.*J. Comp. Neurol.* 1995; 363:642--664.

Carrera E, Michel P, Bogousslavsky J. Anteromedian, central and posterolateral territory infarcts. Three variant types. *Stroke* 2004;35:2826-2834.

Carter, C., Macdonald, A.M., Botvinick, M., Ross, L., Stenger,V.A., Noll, D., Cohen, J.D. Parsing executive processes: strategic vs. evaluative functions of the anterior cingulated cortex. *Proc. Natl. Acad. Sci. U.S.A.* 2000; 97 (4): 1944–1948.

Castaigne P, Lhermitte F, Buge A, et al. Paramedian thalamic and midbrain infarcts: clinical and neuropathological study. *Ann. Neurol.* 1981;10:127-148.

Ceballos-Baumann AO, Boecker H, Bartenstein P, von Falkenhayn I, Riescher H, Conrad B et al. A positron emission tomographic study of subthalamic nucleus stimulation in parkinson disease: enhanced movement-related activity of motor association cortex and decreased motor cortex resting activity. *Arch. Neurol.* 1999; 56: 997-1003.

Chari G., Shaw PJ., Sahgal A. Non-verbal visual attention but not recognition memory of learning processes are impaired in motor neuron disease. *Neuropsychologia* 1996; 34(5): 377-385.

Chatterjee A, Yapundich R, Mennemeier M, et al. Thalamic thought disorder: on being "a bit addled". *Cortex* 1997;33:419-440.

Chen W, Zhu XH, Thulborn KR, Ugurbil K. Retinotopic mapping of lateral geniculate nucleus in humans using functional magnetic resonance imaging. *Proc. Natl. Acad. Sci. USA* 1999; 96:2430–2434

Chen W, Zhu XH. Correlation of activation sizes between lateral geniculate nucleus and primary visual cortex in humans. *Magn. Reson. Med.* 2001; 45:202–205

Churchill L, Kalivas PW. The involvement of the mediodorsal nucleus of the thalamus and the midbrain extrapyramidal area in locomotion elicited from the ventral pallidum. *Behav. Brain Res.* 1999; 104:63–71

Clarke S, Assal G, Bogousslavsky J, et al. Pure amnesia after unilateral left polar thalamic infarct: topographic and sequential neuropsychological and metabolic (PET) correlations. *J. Neurol. Neurosurg. Psychiatry* 1994;57:27-34.

Collie DA, Summers DM, Sellar RJ, Ironside JW, Cooper S, Zeidler M, Knight R, Will RG.Diagnosing variant Creutzfeldt-Jakob disease with the pulvinar sign: MR imaging findings in 86 neuropathologically confirmed cases. *AJNR Am. J. Neuroradiol.* 2003;24(8):1560-9.

Contreras D., Steriade M. Synchronization of low-frequency rhythms in corticothalamic networks. *Neuroscience* 1997; 76: 11-24.

Copland D. The basal ganglia and semantic engagement: potential insights from semantic priming in individuals with subcortical vascular lesions, Parkinson's disease, and cortical lesions. *J. Int. Neuropsychol. Soc.* 2003; 9: 1041–1052.

Corbett A., Bennet H., Kos S. Cognitive dysfunction following subcortical infarction. *Arch. Neurol.* 1994; 51: 999-1007.

Cosson A, Tatu L, Vuillier F, et al. Arterial vascularization of the human thalamus: extra-parenchymal arterial groups. *Surg. Radiol. Anat.* 2003;25:408-415.

Cox CL, Huguenard JR, Prince DA. Nucleus reticularis neurons mediate diverse inhibitory effects in thalamus. *Proc. Natl. Acad. Sci. USA*. 1997; 94:8854–8859.

Craig AD, Bushnell MC, Zhang ET, Blomqvist A. A thalamic nucleus specific for pain and temperature sensation. *Nature* 1994; 372:770–773

Critchley HD, Rolls ET Hunger and satiety modify the responses of olfactory and visual neurons in the primate orbitofrontal cortex. *J. Neurophysiol.* 1996; 75:1673--1686.

Crosson B, Beenfield H, Allison Cato M, Sadek JR, Moore BAcon A, Wierenga CE, Gopinath K, Soltysik D, Bauer RM, Auerbach EJ, Gokcay D, Leonard CM, Briggs RW. Left and right basal ganglia and frontal activity during language generation: cotributions to lexical, semantic, and phonological processes. *J. Intern. Neuropsych. Soc.* 2003; 9: 1061-1077.

Crosson B. Subcortical functions in language and memory. New York: Guilford. 1992.

Crosson B. Subcortical mechanisms in Language: Lexical-Semantic Mechanisms and the thalamus. *Brain and Cognition* 1999;40:414-438.

Crosson B., Zawacki T., Brinson G., Lu L., Sadek J. Models of subcortical functions in language: current status. *Journal of Neurolinguistics*, 1997; 4: 277-300.

Crosson, B., Sadek, J.R., Bobholz, J.A., Gokcay, D., Mohr, C.M., Leonard, C.M., Maron, L., Auerbach, E.J., Browd, S.R., Freeman, A.J., Briggs, R.W. Activity in the paracingulate and cingulate sulci duringword generation: an fMRI study of functional anatomy. *Cereb. Cortex* 1999; 9 (4): 307–316.

Cummings J.L. Anatomic and behavioural aspects of frontal-subcortical circuits. In: Grafman J., Holyoak KJ, Boller F (Eds) Structure and function of the human prefrontal cortex. *Ann. NY Acad. Sci.* 1995; 769: 1-14.

Cummings JL. Cholinesterase inhibitors for treatment of dementia associated with Parkinson's disease. *J. Neurol. Neurosurg. Psychiatry* 2005; 76: 903–4.

Cummings JL. Frontotemporal dementias: an update. American Academy of Neurology, (AAN), San Diego, 2000; 2FC.002: 72-94.

Cummings JL., Mega M., Gray K., Rosenberg-Thompson S., Carusi DA., Gornbein J. The Neurpsychiatric Inventory: comprehensive assessment of psychopathology in dementia. *Neurology* 1994; 44: 2308-2314.

Cummings, J. L. Anatomical and behavioral aspects of frontal-subcortical circuits. Annals of. New York Academy Science, 1995, 769, 1-13.

Czernecki V, Pillon B, Houeto JL, Pochon JB, Levy R, Dubois B. Motivation, reward, and Parkinson's disease: influence of dopatherapy. *Neuropsychologia*. 2002; 40:2257--2267.

D'Antona R, Baron JC, Samson Y, Serdaru M, Viader F, Agid Y, Cambier J. Subcortical dementia: frontal cortex hypometabolism detected by positron tomography in patients with progressive supranuclear palsy. *Brain.* 1985; 108:785—799

Dagher A. Functional imaging in Parkinson's disease. Semin Neurol. 2001;21(1):23-32.

Damasio A.R. Descartes' error: emotion, reason and the human brain. New York : Putnam. 1994.

Damasio AR, Damasio H, Rizzo M, Varney N, Gersh F. Aphasia with nonhemorrhagic lesions in the basal ganglia and internal capsule. *Arch. Neurol.* 1982; 39: 15–24.

Damasio H., Damasio A. Lesion analysis in neuropsychology. New York: Oxford Univer. Press. 1989.

Damasio, A.R., Anderson, S.W. The frontal lobes. In: Heilman, K.M., Valenstein, E. (Eds.), Clinical Neuropsychology. Oxford University Press, New York, 1993. Pp. 409–460.

Das M, Misra UK, Kalita J. A study of clinical, MRI and multimodality evoked potentials in neurologic Wilson disease. *Eur. J. Neurol.* 2007 May;14(5):498-504.

DEEP BRAIN STIMULATION FOR PARKINSON'S DISEASE STUDY GROUP. Deep Brain Stimulation of the Subthalamic nucleus or the pars interna of the globus pallidus in Parkinson's Disease. *New Engl. J. Med.* 2001; 345: 956-963.

Del Tredici K, Rub U, De Vos RA, Bohl JR, Braak H. Where does parkinson disease pathology begin in the brain? *Journal of Neuropathology and Experimental Neurology.* 2002;61(5):413-426.

Delgado M.R., Nystrom L.E., Fissell C., Noll DC., Fiez J.A Tracking the hemodynamic responses to reward and punishment in the striatum. *J. Neurophysiol.* 2000; 84, 3072-3077.

Denis P, Nordmann JP, Elena PP, Lapalus Ph. Autoradiographic localization of retinal dopaminergic receptors in man. *Neurosci. Lett.* 1990; 116:81-6.

Desmond JE, Gabrieli JDE, Glover GH. Dissociation of frontal and cerebellar activity in a cognitive task: evidence for a distinction between selection and search. *Neuroimage* 1998; 7: 368-378.

Destexhe A, Contreras D., Steriade M. Mechanisms underlying the synchronizing action of corticothalamic feedback through inhibition ofthalamic relay cells. *J. Neurophysiol.* 1998; 999-1016.

Destexhe A., Contreras D., Steriade M. Cortically-induced coherence of a thalamic-generated oscillation. *Neuroscience* 1999; 92: 427-443.

Dolan RJ, Paulesu E, Fletcher P. Human memory systems. In Frackowiak RSJ, editor. Human brain function. London: Academic Press 1997: 367-404.

Duara R., Lopez-Alberoloa RF, Barker WW, Loewenstein DA., Zatinsky M., Eisdorfer CE., Weinberg GB. A comparison of familial and sporadic Alzheimer's disease. *Neurology* 1993; 43(7): 1377-1382.

Duda JE, Giasson BI, Mabon ME, Miller DC, Golbe LI, Lee VM, et al. Concurrence of alpha-synuclein and tau brain pathology in the Contursi kindred. *Acta Neuropathologica* 2002;104(1):7-11.

Dujardin K., Defebvre L., Krystkowiak P., Blond S., Destée, A. Influence of chronic bilateral stimulation of the subthalamic nucleus on cognitive function in Parkinson's disease. *Journal of Neurology* 2001; 248: 603-611.

Elbaz A, Bower JH, Maraganore DM, McDonnell SK, Peterson BJ, Ahlskog JE, et al. Risk tables for parkinsonism and Parkinson's disease. *Journal of Clinical Epidemiology* 2002;55(1):25-31.

Elbaz A, Bower JH, Peterson BJ, Maraganore DM, McDonnell SK, Ahlskog JE, et al. Survival study of Parkinson disease in Olmsted County, Minnesota. *Archives of Neurology* 2003;60(1):91-96.

Elliott R, Newman JL, Longe OA, Deakin JF Differential response patterns in the striatum and orbitofrontal cortex to financial reward in humans: a parametric functional magnetic resonance imaging study. *J. Neurosci.* 2003; 23:303--307.

Elliott R., Friston K.J., Dolan R.J. Dissociable neural responses in human reward systems. *J. Neurosci.* 2000; 20: 6159-6165.

Engelborghs S, Marien P, Pickut BA et al. Loss of psychic self-activation after paramedian bithalamic infarction. *Stroke* 2000;31:1762-1765.

Eriksson, S.H., Rugg-Gunn, F.J., Symms, M.R., Barker, G.J., Duncan, J.S. Diffusion tensor imaging in patients with epilepsy and malformations of cortical development. *Brain* 2001; 124: 617–626.

Erkinjuntti T., Ketonen L., Sulkava R., Vuorialho M., Palo J. CT in the differential diagnosis between Alzheimer's disease and vascular dementia. *Acta Neurologica Scandinavica* 1987; 75: 262–270.

Eslinger PJ, Damasio AR Severe disturbance of higher cognition after bilateral frontal lobe ablation: patient EVR. *Neurology* 1985; 35:1731--1741.

Eslinger PJ, Warner GC, Grattan LM, Easton JD. „Frontal lobe" utilization behavior associated with paramedian thalamic infarction. *Neurology* 1991;41:450-452.

Fabbro F. Subcortical aphasia. *Journal of Neurolinguistics*, 1997; 10: 251-367.

Fabbro F., Clarici A., Bava A. Effects of left basal ganglia lesions on language production. *Perceptual and Motor Skills*, 1996; 82: 1291-1298

Fabbro F., Peru A., Skrap M. Language disorders in bilingual patients after thalamic lesions. *Journal of Neurolinguistics* 1997; 10: 347-367.

Fahn S, Quinn N. Unusual movement disorders. American Academy of Neurology, San Diego, 2000; 3AG.003-1-24.

Fahn S., Elton R.L., THE MEMBERS OF THE UPDRS DEVELOPMENT COMMITTEE The Unified Parkinson's disease rating scale. In: S. Fahn, C. D. Marsden, D. B. Calne, M. Goldstein (Eds) Recent developments in Parkinson's disease. Florham Park, NJ: *Macmillan Healthcare*. 1987. Vol. 2. Pp. 153-163.

Fiez J.A., Raife E.A., Balota D.A., Schwarz I.P., Raichle M.B., Petersen S.E. A positron emission tomography study of the short term maintenance of verbal information. *J. Neurosci.* 1996; 16: 808-822.

Fiorillo CD, Tobler PN, Schultz W. Discrete coding of reward probability and uncertainty by dopamine neurons. *Science.* 2003; 299:1898--1902.

Folstein M., Folstein S., McHugh P. Mini-Mental State. A practical method for grading the cognitive state of patients for the clinician. *J. Psych. Res*. 1975; 12: 189-198.

Foreman RD. Mechanism of cardiac pain. *Annu. Rev. Physiol.* 1999; 61:143–147

Fromm, D., Holland, A.L., Swindell, C.S., Reinmuth, O.M.Various consequences of subcortical stroke: prospective study of 16 consecutive cases. *Arch. Neurol.* 1985; 42: 943–950.

Fukatsu R, Fujii T, Yamadori A, et al. Persisting childish behavior after bilateral thalamic infarcts. *Eur. Neurol.* 1997;37:230-235.

Fuster J.M. The Prefrontal Cortex, Raven Press, 1997, New York

Gabrieli, J.D., Poldrack, R.A., Desmond, J.E. The role of left prefrontal cortex in language and memory. *Proc. Natl. Acad. Sci. U.S.A.* 95 1998; (3): 906–913.

Gelb A. Zur medizinischen Psychologie und philosophischen Anthropologie. *Acta Psychologica* 1937; 3: 193-271.

Ghika-Schmid F, Bogousslavsky J. The acute behavioual syndrome of anterior thalamic infarction: a prospective study of 12 cases. *Ann. Neurol.* 2000;48:220-227.

Gilbertson MW, Shenton ME, Ciszewski A, Kasai K, Lasko NB, Orr SP, Pitman RK. Smaller hippocampal volume predicts pathologic vulnerability to psychological trauma. *Nat. Neurosci.* 2002;5(11):1242-7.

Glass M, Dragunow M, Faull RL. Cannabinoid receptors in the human brain: a detailed anatomical and quantitative autoradiographic study in the fetal, neonatal and adult human brain. *Neuroscience* 1997; 77:299–318

Goldman-Rakic P.S, Selemon L.D. New frontiers in basal ganglia research. TINS, 1990; 13: 266-271.

Goldman-Rakic P.S. Cellular and circuit basis of working memory in prefrontal cortex of nonhuman primates. *Prog. Brain Res.*1990; 85: 325-336.

Goldman-Rakic PS. Circuitry of primate prefrontal cortex and regulation of behaviour by representational memory. In: Plum F, Mountcastle U,(Eds): Handbook of physiology, pp. 373--417. Washington, DC: The American Physiological Society. 1987.

Gorham DR. The proverbs test. Psychological Test Specialists, Montana: Missoula. 1956.

Graff-Radford NR, Tranel D, van Hoesen GW. Diencephalic amnesia. *Brain* 1990;11:1-25.

Grasby P.M., Frith C.D., Friston K.J., Bench C., Frackowiak R.S.J., and Dolan R.J. Functional mapping of brain areas implicated in auditory-verbal memory function. *Brain* 1993; 116: 1-2

Graybiel AM, Rauch SL. Toward a neurobiology of OCD. *Neuron* 2000;28(2):343-7.

Greene JG., Smith R., Gardiner M., Timburg GC. Measuring behavioural disturbance of elderly demented in patients in the community and its effects on relatives: a factor analytic study. *Age and Ageing* 1982; 11: 121-126.

Guberman A, Stuss D. The syndrome of bilateral paramedian thalamic infarction. *Neurology* 1983;33:540-546.

Guillery RW. Anatomical evidence concerning the role of the thalamus in corticocortical communication: a brief review. *J. Anat.* 1995; 187:583–592

Gurd J.M, Bamford J.M. Striato-capsular aphasia: contrasting cases. *J. Neuroloinguistics*; 1997; 10(4): 325-346.

Gurd JM, Bessell NJ, Bladon RAW., Bamford JM. A case of foreign accent syndrome, with follow-up clinical, neuropsychological and phonetic descriptions. *Neuropsychologia* 1988; 26: 237-251.

Guridi J, Obeso JA. The subthalamic nucleus, hemiballismus and Parkinson's disease: reappraisal of a neurosurgical dogma. *Brain* 2001; 124: 5-19.

Haber SN The primate basal ganglia: parallel and integrative networks. *J. Chem. Neuroanat.* 2003; 26:317--330.

Haber SN, Goenewegen HJ, Grove EA, et al. Efferent connections of the ventral pallidum in the rat. Evidence of a dual striato-pallidofugal pathways. *J. Comp. Neurol.* 1985;238:322-335.

Haber SN. Integrating cognition and motivation into the Basal ganglia pathways of action. In: MA Bedard et al. (Eds) Mental and behavioural dysfunction in Movement Disorders. Human Press Inc. Totowa NJ, 2003. Pp. 35-50.

Hamilton JM, Salmon DP, Corey-Bloom J, Gamst A, Paulsen JS, Jerkins S, Jacobson MW, Peavy G Behavioural abnormalities contribute to functional decline in Huntington's disease. *J. Neurol. Neurosurg. Psychiatry* 2003; 74:120--122.

Hamilton M. A rating scale for depression. *Journal of Neurology, Neurosurgery, Psychiatry* 1960; 23: 5662-5666.

Hashimoto R, Yoshida M, Tanaka Y. Utilisation behavior after right thalamic infarction. *Eur. Neurol.* 1995;35:58-62.

Haslinger B, Erhard P., Kampfe N, Boecker H, Rummeny E, Schwaiger M, et al. Event-related functional magnetic resonance imaging in Parkinson's Disease before and after levo-dopa. *Brain* 2001; 124: 558-570.

Hassani O.K., Cromwell H.C., Schultz W. Influence of expectation of different rewards on behaviour related neuronal activity in the striatum. *J. Neurophysiol.* 2001; 85: 2477-2489.

Hedreen JC, DeLong MR. Organization of striatopallidal, striatonigral, and nigrostriatal projections in the macaque. *J. Comp. Neurol.* 1991;304:569–595

Herrero MT, Barcia C, Navarro JM Functional anatomy of thalamus and basal ganglia *Child's Nerv. Syst.* 2002; 18:386–404

Herynek V, Babis M, Trunecka P, Filip K, Vymazal J, Dezortová M, Hájek M. Chronic liver disease: relaxometry in the brain after liver transplantation. *MAGMA.* 2001 Mar;12(1):10-15

Hikosaka O. Basal ganglia- possible role in motor coordination and learning. *Current Opinion in Neurobiology* 1991; 1: 638-643.

Hikosaka O., Wurtz RH. Visual and oculomotor functions of monkey substantia nigra pars reticulata. L Relation of visual and auditory responses to saccades. *J. Neurophysiol.*1983; 49: 1230-1253.

Hikosaka O., Sakamoto M, Miyashita N. Effects of caudate nucleus stimulation on substantia nigra cell activity in monkey. *Exp. Brain Res.* 1993;95: 457-472.

Hillis, A.E., Barker, P.B., Wityk, R.J., Aldrich, E.M., Restrepo, L., Breese, E.L.,Work, M.. Variability in subcortical aphasia is due to variable sites of cortical hypoperfusion. *Brain* Lang. 2004; 89: 524–530.

Hillis, A.E., Wityk, R.J., Barker, P.B., Beauchamp, N.J., Gailloud, P., Murphy, K., Cooper, O., Metter, E.J. Subcortical aphasia and neglect in acute stroke: the role of cortical hypoperfusion. *Brain* 2002; 125: 1094–1104.

Hirsch EC, Perier C, Orieux G, Francois C, Feger J, Yelnik J, Vila M, Levy R, Tolosa ES, Marin C, Trinidad HM, Obeso JA, Agid Y. Metabolic effects of nigrostriatal denervation in basal ganglia. *Trends Neurosci.* 2000;23:S78–S85.

Hohen, M.M, Yahr MD. Parkinsonism: onset, progression and mortality. *Neurology.* 1967; 17: 427-442.

Hokoc JN, Mariani AP. Tyrosine hydroxlase immunoreactivity in the rhesus monkey retina reveals synapses from bipolar cells to dopaminergic amacrine cells. *J. Neurosci.* 1987; 7:2785-93.

Hollerman JR, Tremblay L, Schultz W. Influence of reward expectation on behavior-related neuronal activity in primate striatum. *J. Neurophysiol.* 1998; 80:947--963.

Hughes AJ, Daniel SE, Kilford L, Lees AJ. Accuracy of clinical diagnosis of idiopathic Parkinson's disease: a clinico-pathological study of 100 cases. *J. Neurol. Neurosurg. Psychiatry* 1992; 55: 181-184.

Hurtig HI, Trojanowski JQ, Galvin J, Ewbank D, Schmidt ML, Lee VM-Y, et al. Alpha-synuclein cortical Lewy bodies correlate with dementia in Parkinson's disease. *Neurology* 2000;54:1916-1921.

Iansek R, Rosenfeld JV, Huxham FE. Deep brain stimulation of the subthalamic nucleus in Parkinson's disease. *MJA,* 20002: 177 (3): 142-146

Isaacs, E.B., Edmonds, C.J., Lucas, A., Gadian, D.G.. Calculation difficulties in children of very low birthweight: a neural correlate. *Brain* 2001; 124: 1701–1707.

Isella V, Melzi P, Grimaldi M, Iurlaro S, Piolti R, Ferrarese C, Frattola L, Appollonio I. Clinical, neuropsychological, and morphometric correlates of apathy in Parkinson's disease. Mov. Disord. 2002; 17:366-371.

Ishai, A., Ungerleider, L., Martin, A., Schouten, J., Haxby, J. Distributed representation of objects in the human ventral visual pathway. *Proc. Natl. Acad. Sci. U.S.A.* 1999; 96 (16), 9379–9384.

Jahanshahi M, Ardouin CMA., Brown RG., Rothwell JC., Obeso J., Albanese A., Rodriguez-Oroz .MC., Moro E, Benabid AL., Pollak P, Limousin-Dowsey P. The impact of deep brain stimulation on executive function in parkinson's disease. *Brain* 2000; 123, 1142-1154.

Jahanshahi M, Jenkins IH, Brown RG, Marsden CD, Passingham RE, Brooks DJ. Self-initiated versus externally triggered movements. I. An investigation using measurement of regional cerebral blood flow with PET and movement-related potentials in normal and Parkinson's disease subjects. *Brain.* 1995; 118:913--33.

Jan C, Francois C, Yelnik J, Tande D, Agid Y, Hirsch EC. Dopaminergic innervation of the pallidum in the normal state, MPTP-treated monkeys and Parkinsonian patients. *Eur. J. Neurosci.* 2000;12:4525–4535.

Jankovic J Parkinson's disease: clinical features and diagnosis *J. Neurol. Neurosurg. Psychiatry* 2008;79;368-376

Jarraya B, Bonnet A-M, Duyckaerts C, et al. Parkinson's disease, subthalamic stimulation and selection of candidates: a pathologic study. *Mov. Disord.* 2003; 18: 1517–20.

Jenkins IH, Jahanshahi M, Jueptner M, Passingham RE, Brooks DJ. Self-initiated versus externally triggered movements. II. The effect of movement predictability on regional cerebral blood flow. *Brain* 2000; 123:1216--28.

Jones E.G. The thalamus of primates. In: Bloom, F. E., Bjorklund, A., Hokfelt, T. (Eds.) The Primate Nervous System, Part II, VoI. 14, Elsevier Science, Amsterdam, 1998. Pp.1-298

Jones EG. Correlation and revised nomenclature of ventral nuclei in the thalamus of human and monkey. Stereotact Funct Neurosurg. 1990; 54/55:1–20

Jones EG. The thalamus. Plenum, 1985. New York

Jueptner M., Frith C.D., Brooks DL, Frackowiak R.S., Passingham RE. Anatomy of molar learning. Subcortical structures and learning by trial and error. *J. Neurophysiol.* 1997, 77, 1325-1337. A

Jueptner M., Stephan K.M., Frith CD., Brooks D.I., Frackowiak R.S., Passingham R.E. Anatomy of motor learning. Frontal cortex and attention to action. *J. Neurophysiol.*1997; 77: 1313-1324. B

Jueptner M., Weiller C. A review of differences between basal ganglia and cerebellar control of movements as revealed by functional imaging studies. *Brain* 1998; 121:1437–1449.

Kapur N, Coughlan K. Confabulations and frontal lobe dysfunction. *J. Neurol. Neurosurg. Psychiatry* 1980;43:461-463.

Karnath HO, Ferber S, Himmlebach M. Spatial awareness is a function of the temporal not the posterior parietal lobe. *Nature* 2001;411:950-953.

Karnath HO, Himmelbach M, Rorden C. The subcortical anatomy of human spacial neglect: putamen, caudate nucleus and pulvinar. *Brain* 2002;125:350-360.

Kawaguchi Y., Wilson CJ., Augood S.I., Emson P.C Striatal interneurones: chemical, physiological and morphological characterization. *Trends Neurosci.* 1995; 18,527-535.

Kennedy M, Murdoch BE. Chronic aphasia subsequent to striato-capsular and thalamic lesions in the left hemisphere. Brain and Language 1993;44:284-295.

Kertesz A., Munoz DG. Pick's disease and Pick complex. Wiley-Liss: New York. 1998.

Kertesz, A.. Subcortical agraphia. In: Vallar, G., Cappa, S.F., Wallesch, C.W. (Eds.), Neuropsychological Disorders Associated with Subcortical Lesions. Oxford Univ. Press, Oxford. 1992. Pp. 334–356.

King AD, Walshe JM, Kendall BE, Chinn RJ, Paley MN, Wilkinson ID, Halligan S, Hall-Craggs MA. Cranial MR imaging in Wilson's disease. *AJR Am. J. Roentgenol.* 1996 Dec;167(6):1579-84.

Klingberg, T., Hedehus, M., Temple, E., Salz, T., Gabrieli, J.D., Moseley, M.E., Poldrack, R.A. Microstructure of temporo-parietal white matter as a basis for reading ability: evidence from diffusion tensor magnetic resonance imaging. *Neuron* 2000; 25: 493–500.

Klos K J, Ahlskog J E, Kumar N , Cambern S, Butz J, Burritt M , Fealey R D, Cowl C T, Parisi J E, Josephs K A Brain metal concentrations in chronic liver failure patients with pallidal T1 MRI hyperintensity. *Neurology*. 2006 Dec 12;67 (11):1984-9

Klos KJ, Ahlskog JE, Kumar N, Cambern S, Butz J, Burritt M, Fealey RD, Cowl CT, Parisi JE, Josephs KA. Brain metal concentrations in chronic liver failure patients with pallidal T1 MRI hyperintensity. *Neurology.* 2006 Dec 12;67(11):1984-1989.

Kritchevsky M, Graff-Radford NR, Damasio AR. Normal memory after damage to medial thalamus. *Arch. Neurol.* 1987;44:959-962.

Kumar E, Mashih AK, Pardo J. Global aphasia due to thalamic hemorrhage: a case report and review of the literature. *Arch. Phys. Med. Rehabil.* 1996;77:1312-1315.

Kumar R., Lozano A. M., Montgomery E., Lang A. E. Pallidotomy and deep brain stimulation of the pallidum and subthalamic nucleus in advanced Parkinson's disease. *Movement Disorders*, 1998; 13(S1): 73-82.

Kumral E, Evyapan D, Balkir K. Acute caudate vascular lesions. *Stroke* 1999; 30:100—108.

Labauge R, Pages M, Marty-Double C, et al. Occlusion of the basilar artery. A review with 17 personal cases. *Rev. Neurol.* 1981;137:545-571.

Laplane D, Levasseur M, Pillon B, Dubois B, Baulac M, Mazoyer B, Tran Dinh S, Sette G, Danze F, Baron JC. Obsessive-compulsive and other behavioural changes with bilateral basal ganglia lesions. A neuropsychological, magnetic resonance imaging and positron tomography study. *Brain.* 1989; 112 (Pt 3):699—725.

Laplane D. La perte d'auto-activation psychique. *Rev. Neurol.* 1990;146:397-404.

Lapper SR, Bolam JP. Input from the frontal cortex and the parafascicular nucleus to cholinergic interneurons in the dorsal striatum of the rat. *Neuroscience* 1992;51:533–545.

Lasater EM, Dowling JE. Dopamine decreases conductance of the electrical junctions between cultured retinal horizontal cells in carp retina. *Nature* 1983; 301:243-6.

Lawrence AD, Watkins LHA, Sahakian BJ Hodges JR, Robbins TW. Visual object and visuospatial cognition in Huntington's disease: implications for information processing in corticostriatal circuits. *Brain* 2000 ; 123: 1349-1364

Layrargues GP. Movement dysfunction and hepatic encephalopathy. *Metab. Brain Dis.* 2001 Jun;16(1-2):27-35.

Lee AC, Harris JP, Calvert JE. Impairments of mental rotation in Parkinson's disease. *Neuropsychologia* 1998;36:109-114.

Lee EJ, Park JH, Ihn Y, Kim YJ, Lee SK, Park CS.Acute bilateral basal ganglia lesions in diabetic uraemia: diffusion-weighted MRI. *Neuroradiology.* 2007 Dec;49(12):1009-13.

Levin BE, Tomer R, Rey GJ. Cognitive impairments in Parkinson's disease. *Neurol. Clin.* 1991; 10:471-485.

Levy ML., Miller BL., Cummings JL., Fairbanks LA., Craig A. Alzheimer disease and frontotemporal dementias: behavioural distinctions. *Arch. Neurol.* 1996; 53: 687-690.

Levy R, Dubois B. Apathy and the Functional Anatomy of the Prefrontal Cortex--Basal Ganglia Circuits. *Cerebral Cortex* 2006;16:916—928.

Li JY, Yong TY, Sebben R, Khoo E, Disney AP.Bilateral basal ganglia lesions in patients with end-stage diabetic nephropathy. *Nephrology (Carlton).* 2008 Feb;13(1):68-72

Limousin P, Pollack P, Benazzouz A, Hoffmann D, Le Bas JF, Brousolle E et al. Effect of parkinsonian signs and symptoms of bilateral subthalamic nucleus stimulation. *Lancet* 1995; 345: 91-95

Limousin P., Greene J., Pollak P., Rothwell J., Benabid A.L., Frackowiak R. Changes in cerebral activity pattern due to subthalamic nucleus or internal pallidum stimulation in Parkinson's disease. *Annals of Neurology*, 1997, 42, 283-291.

Longworth C. E., Keenan S. E., Barker R. A., Marslen-WilsonW. D, Tyler L. K. The basal ganglia and rule-governed language use: evidence from vascular and degenerative conditions. *Brain* 2005; ,128: 584–596.

Looi JCL., Sachdev PS. Differentiation of vascular dementia from AD on neuropsychological tests. *Neurology* 1999; 53: 670–678.

Lozza C, Marie RM, Baron JC. The metabolic substrates of bradykinesia and tremor in uncomplicated Parkinson's disease. *Neuroimage* 2002;17:688–99.

Luria A. R. Traumatic aphasia. Den Haag: Mouton.1970.

Luria AR. Higher cortical functions in man. Prefaces to the English ed. by Hans-Lukas Teuber and Karl H. Pribram. Authorized translation from the Russian, by Basil Haigh. New York, Basic Books. 1966.

Lynd-Balta E, Haber SN. Primate striatonigral projections: a comparison of the sensorimotor-related striatum and the ventral striatum. *J. Comp. Neurol.* 1994;345:562–578.

MacDonald, A.W., Cohen, J.D., Stenger, V.A., Carter, C.S. Dissociating the role of the dorsolateral prefrontal and anterior cingulate cortex in cognitive control. *Science* 2000; 288 (5472): 1835–1838.

Malamut BL, Graff-Radford NR, Chawluk J, et al. Memory in a case of bilateral thalamic infarction. *Neurology* 1992;42:163-169.

Mangel SC, Dowling JE. Responsiveness and receptive field size of carp horizontal cells are reduced by prolonged darkness and dopamine. *Science* 1985; 229:1107-09.

Manos PJ. The utility of the ten-point clock test as a screen for cognitive impairment in general hospital patients. *General Hospital Psychiatry*, 1997; 8: 469-476.

Maria L. Zeneroli Magnetic Resonance Imaging Vol 9 pp 295-302, 1991

Marin RS Apathy: concept, syndrome, neural mechanisms, and treatment. *Semin. Clin. Neuropsychiatry*. 1996; 1:304—314.

Marin RS Apathy: a neuropsychiatric syndrome. *J. Neuropsychiatry Clin. Neurosci.* 1991; 3:243--254.

Marsden C.D., Obeso J.A. The functions of the basal ganglia and the parodox of stereotaxic surgery in Parkinson's disease. *Brain* 1994; 117: 877-897.

Matison R., Mayeux R., Rosen R., Fahn S. Tip of the tongue phenomenon in Parkinson disease. *Neurology* 1982; 35: 567-570.

Maurice N, Deniau JM, Glowinski J, Thierry AM Relationships between the prefrontal cortex and the basal ganglia in the rat: physiology of the corticosubthalamic circuits. *J. Neurosci*. 1998; 18: 9539-9546.

McCarthy R.A., Warrington E. K. Cognitive neuropsychology: a clinical introduction. Orlando, FL: Academic Press.1990.

McFarland N.R., Haber S.N. Thalamic relay nuclei of the basal ganglia form both reciprocal and nonreciprocal cortical connections linking multiple frontal cortical areas. *J. Neurosci.*2002; 22: 8117-8132.

McGilchrist I, Goldstein LH, Jadresic D, Fenwick P. Thalamo-frontal psychosis. *Br. J. Psychiatry* 1993;163:113-115.

McGonigle P, Wax MB, Molinoff PB. Characterization of binding sites for 3H-spiroperidol in human retina. *Invest. Ophthalmol. Vis. Sci.* 1988;29:687-694.

McPherson SE., Cummings JL. Neuropsychological aspects of vascular dementia. *Brain Cognition* 1996; 31: 261–282.

Menard MT, Kosslyn SM, Thompson WL, Alert NM, Rauch SL. Encoding words and pictures: a positron emission tomography study. *Neuropsychologia* 1996; 34: 185-194.

Mendez MF., Perryman KM., Miller BL., Cummings JL. Behavioral differences between frontotemporal dementia and Alzheimer's Disease: a comparison on the BEHAVE-AD rating scale. *Intern. Journ. Psychoger.*1998; 10: 155-162.

Mendez MF., Selwood A., Mastri AR., Frey WH. Pick's disease versus Alzheimer's disease: a comparison of clinical characteristics. *Neurology* 1993; 43: 289-292.

Mennemeier M, Fennel E, Valenstein E, Heilman KL. Contributions of the left intralaminar and medial thalamic nuclei to memory. *Arch. Neurol.* 1992;2004:1050-1058.

Mestre TA, Ferreira JJ, Pimentel J Putaminal petechial haemorrhage as the cause of non-ketotic hyperglycaemic chorea:a neuropathological case correlated with MRI findings. *J. Neurol. Neurosurg. Psychiatry*. 2007 May ;78 (5):549-50

Metter E., Riege WH, Hanson WR, Jackson CA, Kempler D, Van Lancker D. Subcortical structures in aphasia. An analysis based on (F-18)-fluorodeoxyglucose, positron emission tomography, and computed tomography. Archives of Neurology 1988; 45: 1229-1234.

Middleton FA, Strick PL. Basal ganglia and cerebellar loops: motor and cognitive circuits. *Brain Res. Rev.* 2000; 31 (2-3): 236-250.

Middleton FA. Fundamental and Clinical Evidence for Basal Ganglia Influences on Cognition In: M-A. Bédard et al (Eds) Mental and Behavioral Dysfunction in Movement Disorders Humana Press Inc., Totowa, NJ. 2003.

Miguel EC, Rauch SL, Leckman JF, eds. Neuropsychiatry of the Basal Ganglia. Psychiatric Clinics of North America. Philadelphia: W.B. Saunders, 1997.

Miller BL. Frontotemporal dementia: clinical and behavioral features. American Academy of Neurology (AAN) San Diego, 2000; 3PC.007: 11-22.

Miller BL. Pick's Disease and frontotemporal dementias: clinical and biological overview. American Academy of Neurology (AAN) Philadelphia, 2001; 2FC.005: 123-134.

Miller BL., Ikonte C., Ponton M., Levy M., Boone K., Darby A., Berman N., Mena I., Cummings JL. A study of the Lund- Manchester research criteria for frontotemporal dementia: Clinical and single photon emission CT correlations. *Neurology* 1997; 48: 937-942.

Milner B. Effects of different brain lesions on card sorting. *Arch. Neurol.*1963; 9: 90-100.

Milton WJ, Atlas SW, Lavi E, Mollman JE.Magnetic resonance imaging of Creutzfeldt-Jacob disease. Ann Neurol. 1991 Apr;29(4):438-40.

Mink, J.W. The basal ganglia: focused selection and inhibition of competing motor programs. *Prog. Neurobiol.* 1996; 50 (4): 381–425.

Mishkin M., Manning FI. Non-spatial memory after selective prefrontal lesions in monkeys. *Brain Res.* 1978; 143: 313-323.

Mohr J., Wattas W., Duncan G. Thalamic hemorrhage and aphasia. Brain and Language, 1975; 2: 3-17.

Moreaud O, Charnallet A, David D, et al. Frontal lobe syndrome after a left genu capsular infarction. *Eur. Neurol.* 1996;36:322-324.

Moreaud O, Fournet N, Roulin JL, Naegele B, Pellat J. The phonological loop in medicated patients with Parkinson's disease: presence of phonological similarity and word length effects. *J. Neurol. Neurosurg. Psychiatry.* 1997;62:609-611.

Morecraft RJ, Louie JL, Herrick JL, Stilwell-Morecraft KS. Cortical innervation of the facial nucleus in the non-human primate: a new interpretation of the effects of stroke and related subtotal brain trauma on the muscles of facial expression. *Brain* 2001; 124: 176 – 208.

Moretti R, Torre P, Antonello R M.,Capus L, Gioulis M, Zambito Marsala S,Cazzato G, Bava A. "Speech start-hesitation" following subthalamic nucleus stimulation in Parkinson Disease. *European Neurology*, 2003; 49: 251-253.- B

Moretti R, Torre P., Antonello RM, Capus L, Zambito Marsala S, Cattaruzza T, Cazzato G., Bava A. Neuropsychological changes after subthalamic nucleus stimulation: a 12-month follow-up in nine patients with Parkinson Disease. *Parkinsonism and Related disorders*, 2003; 10: 73-79. A

Moretti R, Torre P., Antonello RM, Cattaruzza T, Cazzato G, Bava A.. Frontal lobe dementia and subcortical vascular dementia: a neuropsychological comparison. *Drugs and Aging*, 2004; 21 (14): 931-937.

Moretti R, Torre P., Antonello RM, Cazzato G., Bava A. Subcortical-cortical lesions and two-step aplasia in a bilingual patient. In: S.P.Shohov (Ed). *Advances in psychology research.* Vol. 23. Novascience Publisher, New York, 2003. Pp. 33-44. C

Moretti R, Torre P., Antonello RM., Capus L, Gioulis M, Zambito Marsala S., Cazzato G, Bava A. Cognitive changes following subthalamic nucleus stimulation in two patients with parkinson disease. *Perceptual and motor skills*, 2002, 95: 477-486.

Moretti R., Torre P. La demenza frontale. Aprilia-Roma: Stampa Grafica-2000. 2003.

Moretti R., Torre P., Antonello R.M., Capus L., Gioulis M., Zambito Marsala, S., Cazzato G, Bava A. Effects on cognitive abilities following subthalamic nucleus stimulation in Parkinson Disease. *European Journal of Neurology* 2001; 8: 726-727.

Morris R, Petrides M, Pandya DN. Architecture and connections of retrosplenial area 30 in the rhesus monkey (Macaca mulatta). *Eur. J. Neurosci.* 1999; 11: 2506-2518.

Morrow TJ, Casey KL. State related modulation of thalamic somatosensory responses in the awake monkey. *J. Neurophysiol.* 1992; 67:305–317.

Motomura N, Yamadori A, Mori E, et al. Unilateral spatial neglect due to hemorrhage in the thalamic region. *Acta Neurol. Scand.* 1986;74:190-194.

Müller A, Baumgartner RW, Röhrenbach C, Regard M. Persistent Klüver-Bucy syndrome after bilateral thalamic infarction. *NNBN* 1999;12:136-139.

Müller U, von Cramon Y, Pollmann S. D1- versus D2 receptor modulation of visuospatial working memory in humans. *J. Neurosci.* 1998; 18:2720-2728.

Munoz DG., Kertesz A. Frontotemporal lobar atrophies: the Pick complex. In: Qizilbash N., Schneider LS., Chui H., Tariot P., Brodaty H., Kaye J., Erkinjuntti T. Evidence-based dementia practice. Oxford: Blackwell Publishing. 2003. Pp. 297-311.

Nadeau, S.E., Crosson, B. Subcortical aphasia. *Brain Lang.* 1997; 158: 355–402.

Naeser, M.A., Alexander, M.P., Helm-Estabrooks, E.N., Levine, H.L., Laughlin, S.A., Geschwind, N., 1982. Aphasia with predominantly subcortical lesion sites: description of three capsular/putamal aphasia syndromes. *Arch. Neurol.* 1982; 39: 2–14.

Naeser, M.A., Martin, P.I., Nicholas, M., Baker, E.H., Seekins, H., Kobayashi, M., Theoret, H., Fregni, F., Maria-Tormos, H., Kurland, J., Doran, K.W., Pascual-Leone, A. Improved picture naming in chronic aphasia after TMS to part of right Broca's area: an open-protocol study. *Brain Lang.* 2005; 93: 95–105.

Nauta WJH, Mehler WR. Projections of the lentiform nucleus in the monkey. *Brain Res.* 1966;1:3–42.

Nauta WJH. Neural associations of the amygdaloid complex in the monkey. *Brain* 1962;85:505-520.

Nauta WJH. Some efferent connections of the prefrontal cortex in the monkey. In: Waren JM and Akert K (eds.), The Frontal Granular Cortex and Behavior. McGraw-Hill, New York. 1964. Pp. 397-409.

Neau JP, Bogousslavsky J. The syndrome of posterior choroidal artery territory infarction. *Ann. Neurol.* 1996;39:779-788.

Neville H., Mills D., Lawson D. Fractioning language: different subsystems with different sensitive periods. *Cerebral Cortex* 1992; 2: 244-258.

Nishino H., Ono T., Fukuda M., Sasaki K., Muramoto K.I. Single unit activity in monkey caudate nucleus during operant bar pressing feeding behavior. *Neurosci. Lett.*1981; 21: 105-110.

O'Doherty J, Critchley H, Deichmann R, Dolan RJ Dissociating valence of outcome from behavioural control in human orbital and ventral prefrontal cortices. *J. Neurosi.* 2003; 23:7931--7939.

O'Doherty J, Kringelbach ML, Rolls ET, Hornak J, Andrews C Abstract reward and punishment representations in the human orbitofrontal cortex. *Nat. Neurosci.* 2001; 4:95--102.

O'Doherty J, Rolls ET, Francis S, Bowtell R, McGlone F, Kobal G, Renner B, Ahne G. Sensory-specific satiety-related olfactory activation of the human orbitofrontal cortex. *Neuroreport.* 2000; 11: 893--897.

O'Carroll RE. Neuropsychological aspects of liver disease and its treatment. *Neurochem. Res.* 2008 Apr;33(4):683-90.

O'donnell P., Buxton P J., Pitkin A., Jarvis l. J. The magnetic resonance imaging appearances of the brain in acute carbon monoxide poisoning *clinical radiology* 2000; 55, ( 4): 273-280.

Ogawa K, Toita T, Kakinohana Y, Tominaga D, Miyagi K, Yoshii Y, Horikawa A, Sawada S. A patient with improvement in short-term memory disturbance brought about by radiation therapy for germinoma involving Papez circuit. *Radiat. Med.* 1999; 17:317–322

Oh SH, Lee KY, Im JH, Lee MS. Chorea associated with non-ketotic hyperglycemia and hyperintensity basal ganglia lesion on T1-weighted brain MRI study: a meta-analysis of 53 cases including four present cases. *J. Neurol. Sci.* 2002 Aug 15;200(1-2):57-62.

Ojemann G.A. Models of the brain organization for higher integrative functions derived by electrical stimulation techniques. *Human Neurobiology*, 1982; 1,:243-249.

Olsen TS, Bruhn P, Oberg G. Cause of aphasia in patients with subcortical lesions. *Acta Neurologica Scandinavica* 1984; 69 (98): 311-312.

Ongur D, Price JL. The organization of networks within the orbital and medial prefrontal cortex of rats, monkeys and humans. *Cereb. Cortex.* 2000; 10:206–219.

Osborn, Brain

Ott BR., Lafleche G., Whelihan WM., Buongiorno GW., Albert MS., Fo BS. Impaired awareness of deficits in Alzheimer's Disease. Alzh. Dis. Assoc. Disord.1996; 10: 68-76.

Owen A.M., Doyon J., Dagher A, Sadikot A, Evans A.C. Abnormal basal ganglia outflow in Parkinson' s disease identified with PET. Implications far higher cortical functions. *Brain* 1998;. 121: 949-965.

Owen A.M., Doyon J., Petrides M., Evans A.C. Planning and spatial working memory: a positron emission tomography study in humans. *Eur. J. Neurosci.*1996; 8: 353-364.

Owen A.M., James M., Leigh P.N., et al. Fronto-striatal cognitive deficits at different stages of Parkinson's disease. *Brain* 1992; 115: 1727-1751.

Owen AM. Cognitive planning in humans: neuropsychological, neuroanatomical and neuropharmacological perspectives. *Prog. Neurobiol.* 1997; 53: 431-450.

Packard MG, Knowlton BJ. Learning and memory functions of the basal ganglia. *Ann. Rev. Neurosci.* 2002; 25: 563-593.

Pakkenberg B. Leucotomized schizophrenics loss neurons in the mediodorsal nucleus. *Neuropathol. Appl. Neurobiol.* 1993; 19:373–380

Pandya DN, Van Hoesen GW, Mesulam MM. Efferent connections of the cingulate gyrus in the rhesus monkey. *Exp. Brain Res.* 1981; 42: 319-330.

Papez JW. A proposed mechanism of emotion. Arch Neurol Psychiatry. 1937; 38:725–743

Paradis M. Bilingual and polyglot aphasia. In F. Boller and K. J. Grafman (Eds.), *Handbook of Neuropsychology*. Amsterdam: Elsevier. 1989. Pp. 117-140.

Paradis M. Neurolinguistic aspects of implicit and explicit memory: implications for bilingualism and second language acquisition. In N. Ellis (Ed.), Implicit and explicit language learning. London: Academic Press. 1994. Pp. 393-419.

Paradis M. The assessment of bilingual aphasia. Hillsdale, NJ: Erlbaum. 1987.

Paradis M., Canzanella M. Test per l'afasia in un bilingue. Versione italiana. Hillsdale, N.J:. 1990. Erlbaum.

Parashos SA, Maraganore DM, O'Brien PC, Rocca WA. Medical services utilization and prognosis in Parkinson disease: a population-based study. *Mayo Clinic Proceedings* 2002;77(9):918-925.

Pare, D., Curro' Dossi, R., Steriade, M. Neuronal basis of the parkinsonian resting tremor: a hypothesis and its implications far treatment. *Neuroscience* 1990; 35: 217-226.

Pascual-Leone A., Grafman L, Hallett M. Procedural learning and prefrontal cortex. Ann. *NY Acad. Sci.* 1995; 769: 61-70.

Pasquier F, Lebert F, Lavenu I, Guillaume B. The clinical picture of frontotemporal dementia: diagnosis and follow-up. *Dement. Geriatr. Cogn. Disord.* 1999; 10 Suppl 1:10--14.

Passingham R. The Frontal Lobes and Voluntary Action, Oxford University Press, 1993; Oxford.

Paulesu E., Frith CD., Frackowiak R.S.J. The neural correlates of the verbal component of working memory. *Nature* 1993; 362: 342-345.

Pepin EP, Auray-Pepon L. Selective dorsolateral frontal lobe dysfunction associated with diencephalic amnesia. *Neurology* 1993; 43:733–741

Perani D., Vallar G., Cappa S., Messa C., Fazio F. Aphasia and neglect after subcortical stroke. A clinical/cerebral perfusion correlation study. *Brain*, 1987; 110: 1211–1229.

Percheron G, Francois C, Talbi B, Meder JF, Fenelon G, Yelnik J. The primate motor thalamus analyzed with reference to subcortical afferent territories. *Stereotact. Funct. Neurosurg.* 1991; 93:32–41

Percheron G, François C, Talbi B, Yelnik J, Fénelon G. The primate motor thalamus. *Brain Res. Rev.* 1996, 22:93–181

Percheron G, François C, Yelnik J, Fénelon G, Talbi B. The basal ganglia related system of primates: definition, description and informational analysis. In: Percheron G, Mckenzie JS, Feger J (eds) The basal ganglia IV. Plenum, New York, 1994. Pp 3–20

Percheron G. Les artères du thalamus humain. II. Artères et territoires thalamiques paramédians de l'artère basilaire communicante. *Rev. Neurol.* 1976;132 309-324.

Peterson BS, Skudlarski P, Gatenby JC, Zhang H, Anderson AW, Gore JC. An fMRI study of Stroop word-color interference: evidence for cingulated subregions subserving multiple distributed attentional systems. *Biol. Psychiatry* 1999; 45: 1237-1258

Petrides M, Pandya DN. Dorsolateral prefrontal cortex: comparative cytoarchitectonic analysis in the human and the macaque brain and corticocortical connection patterns. *Eur. J Neurosci.* 1999; 11:1011--1036.

Petrides M. Impairments on nonspatial self-ordered and externally-ordered working memory tasks after lesions of the mid-dorsal part of the lateral frontal cortex in the monkey. *J. Neurosci.* 1995; 15: 359-375.

Petrides, M. and Milner, B. Deficits on subject-ordered tasks after frontal- and temporal-lobe lesions in mans. *Neuropsychologia* 1982; 20: 249-262.

Petrovici JN. Speech disturbances following stereotoaxic surgery in ventrolateral thalamus. *Neurological Review* 1980; 3189-3195.

Piccolino M, Neyton J, Geschenfeld HM. Decrease of gap function permeability induced by dopamine and cyclic AMP in horizontal cell of turtle retina. *J. Neurosci.* 1984; 4:2477-88.

Pillon B, Boller F, Levy R, Dubois B Cognitive deficits in Parkinson's disease. In: Boller F, Grafman J (Eds) Handbook of neuropsychology Amsterdam: *Elsevier*. 2002.

Pillon B., Ardouin C., Damier P., Krack P., Houeto J.L, Klinger H., Bonnet AM., Pollak P., Benabid A. L., Agid Y. Neuropsychological changes between off and on subthalamic nucleus and Globus pallidus stimulation in Parkinson's disease. *Neurology* 2000; 55: 411-418.

Pollak, P. Deep brain stimulation. American Academy of Neurology (AAN), Philadelphia, 2001: 3PC. 005-16-52.

Postle BR, Jonides J, Smith EE, Corkin S, Growden JH. Spatial, but not object, delayed response is impaired in early Parkinson's disease. *Neuropsychology* 1997;11:171-179.

Prockop L.D, Chichkova R.I. Carbon monoxide intoxication: An updated review *Journal of the Neurological Sciences*, 2007; 262 (1): 122-130.

Raeva S. Localization in human thalamus of units triggered during "verbal commands", voluntary movements and tremor. *Electroencephalogr. Clin. Neurophysiol.* 1986; 63:160–173.

Rahman S., Sahakian BJ., Hodges JR., Rogers RD., Robbins TW. Specific cognitive deficits in mild frontal variant fronttemporal dementia. *Brain* 1999; 122 (8): 1469-1493.

Ransmayr G. Difficulties in the clinical diagnosis of vascular dementia and dementia of the Alzheimer type: comparison of clinical classifications. *Journal of Neural Transmission* 1998; 53: 79-90.

Rauch SL. Neuropsychiatry of the limbic system and basal ganglia: Examples from imaging studies of anxiety disorders. American Academy of Neurology (AAN). Miami, 2005. 8AC,004 pdf. 1-12.

Raven J. Guide to the Standard Progressive Matrices. London: H. K. Lewis 1960.

Reisberg B., Borenstein J., Salob SP., Ferris SH., Franssen E., Georgota A. Behavioral symptoms in Alzheiemer's disease: phenomenology and treatment. *Journal of Clinical Psychiatry*, 1987; 48 (S1): 9-15.

Richfield EK, Twyman R, Berent S. Neurological syndrome following bilateral damage to the head of the caudate nuclei. *Ann. Neurol.* 1987; 22:768--771.

Robbins TW, Everitt BJ. Neurobehavioural mechanisms of reward and motivation. *Curr. Opin Neurobiol.* 1996; 6:228--236.

Rolls E, Johnstone S. Neuropsychological analysis of striatal function. In: G. Vallar, Cappa SF, Wallesch CW. (Eds) *Neuropsychological disorders associated with subcortical lesions.* Oxford Associated Press, Oxford. 1992.

Rolls ET The neurophysiological basis of brain-stimulation reward. In: Wauquier A, Rolls ET (Eds). Brain stimulation rewards. 1976. Pp 65--87. Amsterdam: North Holland.

Rolls ET, Hornak J, Wade D, McGrath J. Emotion-related learning in patients with social and emotional changes associated with frontal lobe damage. *J. Neurol. Neurosurg. Psychiatry.* 1994; 57:1518--1524.

Rolls ET, Sienkiewicz ZJ, Yaxley S. Hunger modulates the responses to gustatory stimuli of single neurons in the caudolateral orbitofrontal cortex of the macaque monkey. *Eur. J. Neurosci.* 1989; 1:53--60.

Roman GC., Tatemichi TK., Erkinjuntti T., Cummings JL., Masdeu JC., Garcia JH., Amaducci L., Orgogozo JM., Brun A., Hofman A., Moody DM., O'Brien MD., Yamaguchi T., Grafman G., Drayer BP., Bennett DA., Fisher M., Pajeau AI., Beli MA., De CArli C., Culebras A., Korczyn AD., Bogousslavsky J., Hartmann A., Scheinberg P. Vascu!ar dementia: Diagnostic criteria for research studies. (Report of the NINDS-AIREN Intemational Workshop) *Neurology* 1993; 43: 250-260.

Rosene DL, van Hoesen GW. Hippocampal efferents reach widspread areas of cerebral cortex and amygdala in the rhesus monkey. *Science* 1981; 98:315-317.

Rosenkilde C.E. Functional heterogeneity of the prefrontal cortex in the monkey: a review. Behav. *Neural. Biol.* 1979; 25: 301-345.

Rowan A, Vargha-Khadem F, Calamante F,Tournier JD, Kirkham FJ, Chong WK, Baldeweg T, Connelly A, Gadian DG. Cortical abnormalities and language function in young patients with basal ganglia stroke NeuroImage 2007; 36: 431–440.

Ruberg M, Agid Y. Dementia in Parkinson's disease. In: Iversen L, Iversen SD, Snyder SH. *Handbook of psychopharmacology* (Eds), pp. 157--205. New York: Plenum Press.1988.

Rushworth M.F., Nixon P.D., Eacott M.I., Passingham R.E. Ventral prefrontal cortex is not essential for working memory. *J. Neurosurg.* 1997; 17: 4829-4838.

Russchen FT, Bakst I, Amaral DG, Price JL. The amygdalostriatal projections in the monkey. An anterograde tracing study. *Brain Res.* 1985;329:241–257.

Sabatini U, Boulanouar K, Fabre N, Martin F, Carel C, Colonnese C et al.. Cortical motor reorganization in akinetic patients with Parkinson's Disease: a functional MRI study. *Brain* 2000; 123: 394-403.

Saez de Ocariz M, Nader JA, Santos JA, Bautista M. Thalamic vascular lesions. Risk factors and clinical course for infarcts and hemorrhages. *Stroke* 1996;27:1530-1536.

Saint Cyr JA, Trepanier LL, Kumar R, Lozano AM, Lang AE. Neuropsychological consequences of chronic bilateral stimulation of the subthalamic nucleus in Parkinson's Disease. *Brain* 2000; 123: 2091-2108.

Saint-Cyr J.A, Taylor A.E, Trepanier L.L, Lang A.E. The caudate nucleus: head ganglion of the habit system. In: G. Vallar, Cappa SF, Wallesch CW. (Eds) Neuropsychological disorders associated with subcortical lesions. Oxford Associated Press, Oxford. 1992.

Saint-Cyr JA, Ungerleider LG, Desimone R Organization of visual cortical inputs to the striatum and subsequent outputs to the pallidonigral complex in the monkey. *J. Comp. Neurol.* 1990; 298:129-156.

Sakai K., Hikosaka O., Miyauchi S., Takino R., Sasaki Y., Putz B. Transition of brain activation from frontal to parietal areas in visuomotor sequence learning. *J. Neurosci.* 1998; 18: 1827-1840.

Sakai ST, Stepniewska I, Qi HX, Kaas JH. Pallidal and cerebellar afferents to pre-supplementary motor area thalamocortical neurons in the owl monkey: a multiple labelling study. *J. Comp. Neurol.* 2000; 417:164–180

Samuel M, Ceballos-Baumann AO, Turjanski N, Boecker H, Gorospe A, Linazasoro G, Holmes AP, DeLong MR, Vitek JL, Thomas DG, Quinn NP, Obeso JA, Brooks DJ. Pallidotomy in Parkinson's disease increases supplementary motor area and prefrontal activation during performance of volitional movements an H2(15)O PET study. *Brain* 1997; 120:1301--1313.

Scatton B, Javoy-Agid F, Rouqier L, Dubois B, Agid Y. Reduction of cortical dopamine, noradrenaline, serotonin and their metabolites in Parkinson's disease. *Brain Res.* 1983; 275:321-328

Schenck CH, Bundlie SR, Mahowald MW. Delayed emergence of a parkinsonian disorder in 38% of 29 older men initially diagnosed with idiopathic rapid eye movement sleep behaviour disorder. *Neurology*. 1996;46(2):388-393.

Schmahmann JD, Pandya DN. Anatomical investigation of thalamic projections to the posterior parietal cortices in rhesus monkey. *J. Comp. Neurol.* 1990;295:299-326.

Schmahmann JD., Pandya DN. Cerebral white matter Fiber pathways. American Academy of Neurology (AAN), 2005, San Francisco, 8AC-004: 114-130.

Schnider A, Gutbrod K, Hess CW, Schroth G. Memory without context: amnesia with confabulati*ons after infarction of the right capsular genu.* J. Neurol. Neurosurg. Psychiatry 1996;61:186-193.

Schorderet M, Nowak JZ. Retinal dopamine D1 and D2 receptors: characterization by binding or pharmacological studies and physiological functions. *Cell. Mol. Neurobiol.* 1990; 10:303-25.

Schrag A, Good CD, Miszkiel K, Morris HR, Mathias CJ, Lees AJ, Quinn NP.Differentiation of atypical parkinsonian syndromes with routine MRI. *Neurology*. 2000 Feb 8;54(3):697-702

Schroeder U, Kuehler A, Haslinger B, Erhard P, Fogel W, Tronnier VM, Lange KW, Boecker H, Ceballos-Baumann AO. Subthalamic nucleus stimulation affects striato-anterior cingulated cortex circuit in a response conflict task: a PET study. *Brain* 2002; 125 (9): 1995-2004.

Schubert T, Volkmann J, Muller U, Sturm V, Voges J, Freund HJ, von Cramon DY (2002). Effects of pallidal deep brain stimulation and levodopa treatment on reaction-time performance in Parkinson's disease. *Exp. Brain Res*. 144: 8-16.

Schultz W, Dayan P, Montague PR. A neural substrate of prediction and reward. *Science* 1997; 275:1593--1599.

Schultz W. Predictive reward signal of dopamine neurons. *J. Neurophysiol.* 1998; 80: 1-27.

Sefidbakht S, Rasekhi AR, Kamali K, Borhani Haghighi A, Salooti A, Meshksar A, Abbasi HR, Moghadami M, Nabavizadeh SA. Methanol poisoning: acute MR and CT findings in nine patients. *Neuroradiology*. 2007 May;49(5):427-35.

Selemon LD, Goldman-Rakic PS. Longitudinal topography and interdigitation of corticostriatal projections in the rhesus monkeys. *J. Neurosci*. 1985; 5:776—794.

Shallice T. From neuropsychology to mental structure. Cambridge, UK: Cambridge Univer. Press. 1988.

Sherman, S.M. and Guillery, R.W. Functional organization of thalamocortical relays. *J. Neurophysiol.* 1996; 76: 1367-1395.

Shuvaev V.T, Shefer V.I. Structure of neuronal activity of the caudate nucleus of monkeys during decision-making and the realization of the molar program in different variants of a delayed spatial choice task *Neurosci. Behav. Physiol.* 1995; 25: 63-70.

Signoret J.L., Castaigne P., Lhermitte F. Rediscovery of Leborgne's brain: anatomical description with CT scan. *Brain and Language,* 1984; 22: 303-319.

Sirigu A., Zalla T., Pillon B., Grafman L., Dubois B., Agid Y. Planning and script analysis following prefrontal lobe lesions. *Ann. NY Acad. Sci* 1995; . 769, 277-288.

Smeding H.M.M., van den Munckhof P., Esselink R. A.J., Schmand B., Schuurman P. R., Speelman J. D. Reversible cognitive decline after DBS STN in PD and displacement of electrodes. *Neurology* 2007; 68 (15): 1235-1236.

Spahr L, Burkhard PR, Grötzsch H, Hadengue A. Clinical significance of basal ganglia alterations at brain MRI and 1H MRS in cirrhosis and role in the pathogenesis of hepatic encephalopathy. *Metab. Brain Dis.* 2002 Dec;17(4):399-413

Squire L. R. Memory and brain. New York: Oxford Univer.Press. 1987.

Squire L.R., Knowlton B.J. The medial temporal lobe, the hippocampus and the memory systems of the brain. In M. Gazzaniga (Ed.), The new cognitive neurosciences. Cambridge, MA: MIT Press. Pp. 765-779. 2000

Squire LR, Zola-Morgan S. The medial temporal lobe memory system. *Science* 1991; 253: 1380-1386.

Steinmetz H, Seitz R.J. Functional anatomy of language processing: neuroimaging and the problem of individual variability. *Neuropsychologia* 1997; 29: 1149-1161.

Steriade M. Coherent oscillations and short-term plasticity in corticothalamic networks. Trends Neurosci. 1999; 22: 337-345.

Strick PL. Anatomical analysis of ventrolateral thalamic input to primar motor cortex. *J. Neurophysiol.* 1976;39:1021-1031.

Stroop JR. Studies of interference in serial verbal reactions. *Journal of Experimental Psychology* 1935; 18: 643-662.

Stuss DT, Van Reekum R, Murphy KJ. Differentiation of states and causes of apathy. In: Borod JC(ed.): The Neuropsychology of emotion. Oxford: Oxford University Press. 2000. Pp. 340--363.

Tagliati M, Bodis-Wollner I, Kovanecz I, Stanzione P. Spatial frequency tuning in the monkey retina depends on D2 receptor-linked action of dopamine. *Vision. Res.* 1994; 34:2051-2057.

Talairach J, Bancaud J, Geier S, Bordas-Ferrer M, Bonis A, Szikla G, Rusu M. The cingulate gyrus and human behaviour. *Electroencephalogr. Clin. Neurophysiol.* 1973; 34: 45-52.

Tatemichi TK, Desmond DW, Prohovnik I, et al. Confusion and memory loss from capsular genu infarction. A thalamocortical disconnection syndrome ? *Neurology* 1992;42:1966-1979.

Teranishi T, Negishi K, Kato S. Dopamine modulates S-potential amplitude and dye-coupling between horizontal cells in carp retina. *Nature* 1983; 301:243-6.

THE LUND AND MANCHESTER GROUPS. Clinical and neuropathological criteria for frontotemporal dementia. *J. Neurol. Neurosurg. Psych.* 1994; 57: 416-418.

Thobois S, Mertens P, Guenot M, Hermier M, Mollion H, Bouvard M, Chazot G, Broussolle E, Sindou M. Subthalamic nucleus stimulation in parkinson's disease. Clinical evaluation of 18 patients. *Journal of Neurology*, 2002; 249, 529-534.

Thompson JC, Snowden JS, Craufurd D, Neary D Behaviour in Huntington's disease: dissociating cognition-based and mood-based changes. J. *Neuropsychiatry Clin. Neurosci.* 2002; 14:37--43.

Travis R., Sparks D.L. Unitary response and discrimination learning in the squirrel monkey: the globus pallidus. *Physiol. Behav*. 1968; 3: 187-196.

Traykov L., Baudic S., Thibaudet MC., Rigaud AS., Smagghe A., Boller F. Neuropsychlogical deficit in early subcortical vascular dementia: comparison to Alzheimer's disease. *Dem. Geriatr. Cogn. Disord.* 2002; 14: 26-32.

Troster AI, Fields JA, Wilkinson SB, Pahwa R., Miyawaki E., Lyons K.E., Koller WC. Unilateral pallidal stimulation for Parkinson's disease: neurobehavioural functioning before and 3 months after electrode implanatation. *Neurology* 1997; 49: 1078-1083.

Tulving E. Multiple memory systems and consciousness. *Human Neurobiology* 1987; 6: 667-680.

Turner RS, Grafton ST, McIntosh AR, et al. The functional anatomy of parkinsonian bradykinesia. *Neuroimage* 2003;19:163–79.

Uitti RJ, Ahlskog JE, Maraganore DM, Muenter MD, Atkinson EJ, Cha RH, et al. Levodopa therapy and survival in idiopathic Parkinson's disease: Olmsted County project. *Neurology* 1993;43(10):1918-1926.

Vallar G, Perani D, Cappa S.F, Messa C, Lenzi G.L, Fazio F. Recovery from aphasia and neglect after subcortical stroke. *Journal of Neurology, Neurosurgery and Psychiatry* 1988; 51: 1269-1276.

Valldeoriola F, Tolosa E, Alegret M, Rey MJ, Morsi O, Pilleri M, Rumià J. Cognitive changes in Parkinson's disease during subthalamic stimulation: a clinicopathologic study Journal of Neurology, *Neurosurgery, and Psychiatry* 2006;77:565-566

Van Buren J. Confusion and disturbance of speech from stimulation in vicinity of the head of the caudate nucleus. *Journal of Neurosurgery* 1963; 20: 148-157

Van Domburg P, ten Donkelaar HJ, Notermans S. Akinetic mutism with bithalamic infarction. Neuropsychological correlates. *J. Neurol. Sci.* 1996;139:58-65.

Vercueil L., Benazzouz A., Deransart C., Bressand K., Marescaux C., Depaulis A., Benabid A. L. (1998) High frequency stimulation of the subthalamic nucleus suppresses absence seizures in the rat: comparison with neurotoxic lesions. *Epilepsy Research* 1998; 31(1): 39-46.

Wagner, A.D., Pare-Blagoev, E.J., Clark, J., Poldrack, R.A. Recovering meaning: left prefrontal cortex guides controlled semantic retrieval. Neuron 2001; 31 (2): 329–338.

Wallesch C. W., Papagno C. Subcortical aphasia. In F. Clifford Rose, R. Whurr and M. Wukye (Eds.), Aphasia. London: Whurr. 1988. Pp. 256-287.

Wallesch C.W., Johannsen-Horbach H., Bartels C., Herrmann M., 1997. Mechanisms and misconceptions about subcortical aphasia. *Brain Lang.* 1987; 58: 403–409.

Wallesch CW., Henriksen L., Kornhuber HH., Paulson O.B. Observations on regional cerebral blood flow in cortical and subcortical structures during language production in normal man. *Brain and Language* 1985; 25: 224-233.

Watanabe M. Prefrontal unit activity during delayed conditional Go/No-Go discrimination in the monkey. 1. Relation to the stimulus. *Brain Res.* 1986; 382: 1-14.

Watkins K.E., Vargha-Khadem F., Ashburner J., Passingham R.E., Connelly A., Friston K.J., Frackowiak R.S., Mishkin M., Gadian D.G. MRI analysis of an inherited speech and language disorder: structural brain abnormalities. *Brain* 2002; 125: 465–478.

Watson RT, Valenstein E, Heilman KM. Thalamic neglect: Possible role of the medial thalamus and nucleus reticularis in behavior. *Arch. Neurol.* 1981;38:501-506.

Wechsler D. WAIS- Revised manual. New York: The Psychological Corp 1981.

Wechsler D. A standardized memory scale for clinical use. *Journal of Psychology* 1945; 19: 87-97.

Weiller C., Ringelstein E.B., Reiche W., Thron A., Buell U. The large striatocapsular infarct: a clinical and pathophysiological entity. *Neurology* 1990; 47: 1085–1091.

Weiller C., Willmes K., Reich W., Thron A., Isensee C., Buell U., Ringelstein E.B. The case of aphasia or neglect after striatocapsular infarction. *Brain* 1993; 116: 1509–1525.

Weisman D, Hisama FM, Waxman SG, Blumenfeld H. Going deep to cut the link: Cortical dysconnection syndrome caused by a thalamic lesion. *Neurology* 2004;60;1865-1866.

Weisman DC. Yaari R, Hansen LA, Thal LJ.Density of the Brain, Decline of the Mind: An Atypical Case of Fahr Disease. *Arch. Neurol.* 2007;64(5):756-757.

Weissenborn K, Bokemeyer M, Ahl B, Fischer-Wasels D, Giewekemeyer K, van den Hoff J, Köstler H, Berding G. Functional imaging of the brain in patients with liver cirrhosis. *Metab. Brain Dis.* 2004 Dec;19(3-4):269-280

Werring D.J., Toosy A.T., Clark C.A., Parker G.J., Barker G.J., Miller D.H., Thompson A.J. Diffusion tensor imaging can detect and quantify corticospinal tract degeneration after stroke. *J. Neurol., Neurosurg. Psychiatry* 2000; 69: 269–272.

Whelan BM, Bruce E. Murdoch BE, Theodoros DG, Darnell R, Peter Silburn P, Bruce Hall B. Redefining Functional Models of Basal Ganglia Organization: Role for the Posteroventral Pallidum in Linguistic Processing? *Movement Disorders*, 2004; 19, 11: 1267–1278.

Wierenga C.E. Age-related changes in word retrieval: frontal executive vs. temporal semantic substrates. Dissertation. University of Florida, Gainesville, FL. 2005.

Wierenga CE., Benjamin M, Gopinath K, Perlstein WM, Leonard CM, Gonzalez Rothi LJ,Conway T, Cato MA, Briggs R, Crosson B Age-related changes in word retrieval: Role of bilateral frontal and subcortical networks. *Neurobiology of Aging* 2008; 29: 436–451.

Williams GV, Goldman-Rakic PS. Modulation of memory fields by D1 dopamine receptors in prefrontal cortex. *Nature* 1995; 376:572-575.

Witkovsky P, Stone S. Tranchina D. Photoreceptor to horizontal cell synaptic transfer in Xenopus retina: Modulation by dopamine ligands and a circuit model for interactions of rod and cone inputs. *J. Neurophys* 1989; 62:864-81.

Witt K, Pulkowski U, Herzog J, et al. Deep brain stimulation of the subthalamic nucleus improves cognitive flexibility but impairs response inhibition in Parkinson disease. *Arch. Neurol.* 2004;61:697–700.

Xiao Z, Zhang JX, Wang X, Wu R, Hu X, Weng X, Hai Tan L. Differential Activity in Left Inferior Frontal Gyrus for Pseudowords and Real Words: An Event-Related fMRI Study on Auditory Lexical Decision. *Human Brain Mapping* 2005; 25:212–221.

Yakovlev PI, Locke S. Limbic nuclei of the thalamus and connections of limbic cortex, III: corticocortical connections of the anterior cingulated gyrus, the cingulum, and the subcallosal bundle in monkey. *Arch. Neurol.* 1961;5:34-70.

Yamamoto Y, Georgiadis AL, Chang HM, Caplan LR. Posterior cerebral artery territory infarcts in the New England Medical Center Posterior Circulation Registry. *Arch. Neurol.* 1999;56:824-832.

Yamatani, K., Ono T., Nishijo H., Takaku A.Activity and distribution of learning-related neurons in monkey (Macaca fuscata) prefrontal cortex. *Behav. Neurosci.* 1990; 104: 503-537.

Yelnik J, Francois C, Percheron G, Tande D. A spatial and quantitative study of the striatopallidal connection in the monkey. *Neuroreport* 1996;7:985–988.

Yelnik J.Functional Anatomy of the Basal Ganglia Movement Disorders, 2002; 17, 3: S15–S21

Yeterian EH, Pandya DN. Prefrontostriatal connections in relation to cortical architectonic organization in rhesus monkeys. *J. Comp. Neurol.* 1991; 312:43--67.

Zanini S., Melatini A, Vassallo A., Moretti R., Bava A. Speech and language improvement in Parkinson disease after electrical stimulation of the subthalamic nucleus: a case study. *European Journal of Physiology*, 2000; 440: R35-63.

Zatorre R. On the representation of multiple languages in the brain: old problems and new directions. *Brain and Language* 1989; 36: 127-147.

Zung WWK. A self-rating depression scale. *Archives of General Psychiatry* 1965; 12: 63-70.

# Index

## A

## C

## D

## E

## F



## G

## H



## I

## J



## K



## L

## M

## N

## O



## P

## T

## Y



## Z